Human Factors Considerations in the Design and Evaluation of Moving Map Displays of Ownship on the Airport Surface

Michelle Yeh, Ph.D.
U.S. Department of Transportation
Research and Special Programs Administration
John A. Volpe National Transportation Systems Center
Cambridge, MA 02142

DOT/FAA/AR-04/39
DOT-VNTSC-FAA-04-11

Human Factors Research and
Engineering Division,
Federal Aviation Administration
Washington, D.C 20591

September, 2004

Funded by:
Dr. Tom McCloy, Program Manager
Human Factors Research and Engineering Division

Notice

This document is disseminated under the sponsorship of the Department of Transportation in the interest of information exchange. The United States Government assumes no liability for its contents or use thereof.

Notice

The United States Government does not endorse products or manufacturers. Trade or manufacturers' names appear herein solely because they are considered essential to the objective of this report.

REPORT DOCUMENTATION PAGE		*Form Approved* *OMB No. 0704-0188*

Public reporting burden for this collection of information is estimated to average 1 hour per response, including the time for reviewing instructions, searching existing data sources, gathering and maintaining the data needed, and completing and reviewing the collection of information. Send comments regarding this burden estimate or any other aspect of this collection of information, including suggestions for reducing this burden, to Washington Headquarters Services, Directorate for Information Operations and Reports, 1215 Jefferson Davis Highway, Suite 1204, Arlington, VA 22202-4302, and to the Office of Management and Budget, Paperwork Reduction Project (0704-0188), Washington, DC 20503.

1. AGENCY USE ONLY (Leave blank)	2. REPORT DATE September 2004	3. REPORT TYPE AND DATES COVERED Final Report, September 2004
4. TITLE AND SUBTITLE Human Factors Considerations in the Design and Evaluation of Moving Map Displays of Ownship on the Airport Surface		5. FUNDING NUMBERS AB078/FAE2
6. AUTHOR(S) Michelle Yeh		
7. PERFORMING ORGANIZATION NAME(S) AND ADDRESS(ES) U.S. Department of Transportation John A. Volpe National Transportation Systems Center Research and Special Programs Administration Cambridge, MA 02142-1093		8. PERFORMING ORGANIZATION REPORT NUMBER DOT-VNTSC-FAA-04-11
9. SPONSORING/MONITORING AGENCY NAME(S) AND ADDRESS(ES) U.S. Department of Transportation Federal Aviation Administration Office of Aviation Research, Human Factors Research and Engineering Division 800 Independence Avenue, SW Washington, D.C. 20591 Program Manager: Dr. Tom McCloy		10. SPONSORING/MONITORING AGENCY REPORT NUMBER DOT/FAA/AR-04/39
11. SUPPLEMENTARY NOTES		
12a. DISTRIBUTION/AVAILABILITY STATEMENT This document is available to the public through the National Technical Information Service, Springfield, VA 22161		12b. DISTRIBUTION CODE

13. ABSTRACT (Maximum 200 words)

The Federal Aviation Administration (FAA) has requested human factors guidance to support the new moving map Technical Standard Order (TSO)-C165, *Electronic Map Display Equipment for Graphical Depiction of Aircraft Position*. This document was developed to meet that need and is intended to facilitate the identification and resolution of flight-deck human-factors issues associated with presenting an airport surface moving map that depicts ownship location. The guidance in this document contains FAA regulations, industry documents, and user interface design principles that describe good practices relevant to the design of surface moving map displays. This document applies to *all* surface moving map displays regardless of the display platform. Topics in this document cover general user interface issues, the design of surface moving map display elements, the presentation of traffic, and the usability of functions. Additionally, an industry review describing the efforts by manufacturers and research organizations to develop a moving map display with ownship position is provided in Appendix A, and a summary of guidance is included in Appendix B.

14. SUBJECT TERMS surface moving map, aerodrome moving map display, electronic map display, usability, avionics, human factors, design, evaluation		15. NUMBER OF PAGES 142
		16. PRICE CODE
17. SECURITY CLASSIFICATION OF REPORT Unclassified	18. SECURITY CLASSIFICATION OF THIS PAGE Unclassified	19. SECURITY CLASSIFICATION OF ABSTRACT Unclassified	20. LIMITATION OF ABSTRACT

NSN 7540-01-280-5500

Standard Form 298 (Rev. 2-89)
Prescribed by ANSI Std. 239-18
298-102

METRIC/ENGLISH CONVERSION FACTORS

ENGLISH TO METRIC

LENGTH (APPROXIMATE)
- 1 inch (in) = 2.5 centimeters (cm)
- 1 foot (ft) = 30 centimeters (cm)
- 1 yard (yd) = 0.9 meter (m)
- 1 mile (mi) = 1.6 kilometers (km)

AREA (APPROXIMATE)
- 1 square inch (sq in, in^2) = 6.5 square centimeters (cm^2)
- 1 square foot (sq ft, ft^2) = 0.09 square meter (m^2)
- 1 square yard (sq yd, yd^2) = 0.8 square meter (m^2)
- 1 square mile (sq mi, mi^2) = 2.6 square kilometers (km^2)
- 1 acre = 0.4 hectare (he) = 4,000 square meters (m^2)

MASS - WEIGHT (APPROXIMATE)
- 1 ounce (oz) = 28 grams (gm)
- 1 pound (lb) = 0.45 kilogram (kg)
- 1 short ton = 2,000 pounds (lb) = 0.9 tonne (t)

VOLUME (APPROXIMATE)
- 1 teaspoon (tsp) = 5 milliliters (ml)
- 1 tablespoon (tbsp) = 15 milliliters (ml)
- 1 fluid ounce (fl oz) = 30 milliliters (ml)
- 1 cup (c) = 0.24 liter (l)
- 1 pint (pt) = 0.47 liter (l)
- 1 quart (qt) = 0.96 liter (l)
- 1 gallon (gal) = 3.8 liters (l)
- 1 cubic foot (cu ft, ft^3) = 0.03 cubic meter (m^3)
- 1 cubic yard (cu yd, yd^3) = 0.76 cubic meter (m^3)

TEMPERATURE (EXACT)
[(x-32)(5/9)] °F = y °C

METRIC TO ENGLISH

LENGTH (APPROXIMATE)
- 1 millimeter (mm) = 0.04 inch (in)
- 1 centimeter (cm) = 0.4 inch (in)
- 1 meter (m) = 3.3 feet (ft)
- 1 meter (m) = 1.1 yards (yd)
- 1 kilometer (km) = 0.6 mile (mi)

AREA (APPROXIMATE)
- 1 square centimeter (cm^2) = 0.16 square inch (sq in, in^2)
- 1 square meter (m^2) = 1.2 square yards (sq yd, yd^2)
- 1 square kilometer (km^2) = 0.4 square mile (sq mi, mi^2)
- 10,000 square meters (m^2) = 1 hectare (ha) = 2.5 acres

MASS - WEIGHT (APPROXIMATE)
- 1 gram (gm) = 0.036 ounce (oz)
- 1 kilogram (kg) = 2.2 pounds (lb)
- 1 tonne (t) = 1,000 kilograms (kg) = 1.1 short tons

VOLUME (APPROXIMATE)
- 1 milliliter (ml) = 0.03 fluid ounce (fl oz)
- 1 liter (l) = 2.1 pints (pt)
- 1 liter (l) = 1.06 quarts (qt)
- 1 liter (l) = 0.26 gallon (gal)
- 1 cubic meter (m^3) = 36 cubic feet (cu ft, ft^3)
- 1 cubic meter (m^3) = 1.3 cubic yards (cu yd, yd^3)

TEMPERATURE (EXACT)
[(9/5) y + 32] °C = x °F

QUICK INCH - CENTIMETER LENGTH CONVERSION

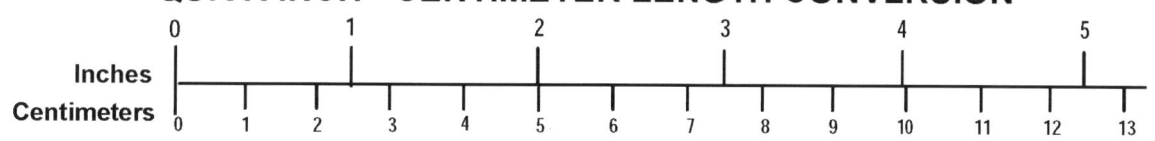

Inches: 0, 1, 2, 3, 4, 5
Centimeters: 0, 1, 2, 3, 4, 5, 6, 7, 8, 9, 10, 11, 12, 13

QUICK FAHRENHEIT - CELSIUS TEMPERATURE CONVERSION

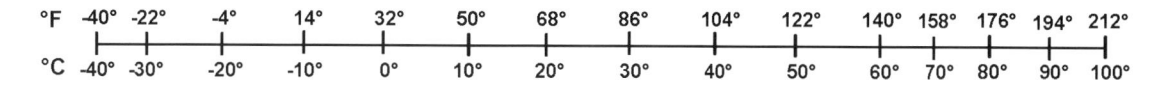

°F: -40° -22° -4° 14° 32° 50° 68° 86° 104° 122° 140° 158° 176° 194° 212°
°C: -40° -30° -20° -10° 0° 10° 20° 30° 40° 50° 60° 70° 80° 90° 100°

For more exact and or other conversion factors, see NIST Miscellaneous Publication 286, Units of Weights and Measures. Price $2.50 SD Catalog No. C13 10286 Updated 6/17/98

TRADEMARK NOTICES

The following information was obtained from company representatives and by searching the database on the United States Patent and Trademark Office website. Other product and company names mentioned herein may be the trademarks of their respective owners. We disclaim liability for errors, omissions or future changes.

ACSS is the registered trademark of Aviation Communication & Surveillance Systems.

Diehl Avionik Systeme is the registered trademark of Diehl Avionik Systeme GmbH.

Garmin AT is a registered trademark of Garmin International, Inc

Honeywell is the registered trademark of Honeywell International Inc.

Jeppesen, FliteDeck, FliteMap, FliteStar, and JeppView are the registered trademarks of Jeppesen Sanderson, Inc.

Rockwell Collins is the registered trademark of Rockwell International Corporation.

MITRE is the registered trademark of The MITRE Corporation.

PREFACE

The Federal Aviation Administration (FAA) has requested human factors guidance to support the new moving map Technical Standard Order (TSO)-C165, *Electronic Map Display Equipment for Graphical Depiction of Aircraft Position*. This document was developed to meet that need. Implementations of the surface moving map function vary widely in the level of detail with which the airport surface is depicted and in the functions that are available. The simplest surface moving map may be ownship position superimposed on a geo-referenced electronic airport diagram. This map may be non-interactive; functionality, if any, will be limited. More complex are surface moving maps constructed from a database that contains positional data describing the location of airport attributes. These displays may vary in the detail with which the airport is depicted and in the functionality available. Common display elements depicted include ownship position, runways, runway labels, taxiways, taxiway labels, non-movement areas, and buildings. Common functions include zooming, displaying traffic, and decluttering. More advanced features such as surface navigation guidance with the display of a taxi route and autozoom are available for some displays.

The FAA specifies requirements for the approval of the surface moving map function in TSO-C165. The TSO references RTCA DO-257A, *Minimum Operational Performance Standards for the Depiction of Navigational Information on Electronic Maps*. This document provides additional guidance to supplement that specified in TSO-C165 and RTCA DO-257A with FAA regulations, industry documents, and general user interface design principles that describe best design practices relevant to the design of the surface moving map displays. The guidance in this document applies to *all* surface moving map displays regardless of the display platform.

This report was prepared by the Operator Performance and Safety Analysis Division of the Office of Research and Analysis at the Volpe Center. It was completed under the Division's Flight Deck Technologies and Procedures Research program. This work is funded by the Human Factors Research and Engineering Division of the FAA and sponsored by the FAA Aircraft Certification Service. Dr. Tom McCloy served as the FAA program manager.

The author would like to thank Divya Chandra, Colleen Donovan, and Bill Kaliardos for reviewing the report and providing valuable feedback. Additionally, the author would like to thank the many industry experts who generously provided information for the industry review.

Feedback on this document can be sent to Michelle Yeh (yeh@volpe.dot.gov).

TABLE OF CONTENTS

List of Tables and Figures	iv
Executive Summary	v
Acronyms	vi
1 Introduction	**1**
1.1 Background	1
1.2 How to Use this Document	2
2 General	**4**
2.1 Use of Color	4
2.2 Alerts and Reminders	7
2.3 Accessibility of Controls	9
2.4 Design of controls	10
2.5 Design of Labels	14
2.6 Control layout	16
2.7 Presentation of Text Information	17
2.8 Symbols	19
2.9 Graphical Icons	21
2.10 Configuring Display Properties	22
2.11 Failure Conditions	23
2.12 Update Rate	25
2.13 Responsiveness	27
2.14 Shared Display Considerations	28
3 Surface-Moving-Map Display Elements	**30**
3.1 Databases	30
3.2 Accuracy	32
3.3 Ownship	37
3.4 Runways	39
3.5 Taxiways	41
3.6 Runway/Taxiway Identifiers	42
3.7 Hold Lines	44
3.8 Non-Movement Areas	45
3.9 Taxi Route	46
3.10 Prioritization of Map Features	48
3.11 Indicators (Velocity Vectors, Compass Rose, Lubber Line)	49
4 Traffic Display	**51**
4.1 Traffic Representations	51
4.2 Selected Traffic	54
4.3 Data Blocks/Data Tags	55
4.4 Altitude Representations	56
5 Functionality	**57**
5.1 Map Range and Panning	57
5.2 Autozoom	59
5.3 Decluttering	60
5.4 Map Orientation	62
References	**64**
Appendix A: Industry Overview	**68**
A.1 Industry Prototypes	**78**
A.1.1 ACSS (Aviation Communication & Surveillance Systems)	78
A.1.2 Diehl Avionik Systeme	80
A.1.3 Garmin Advanced Technologies (AT)	82
A.1.4 Honeywell	86
A.1.5 Jeppesen	88
A.1.6 Rockwell Collins, Inc.	92

A.1.7	Smiths Aerospace	98
A.2	**Research Prototypes**	**100**
A.2.1	William J. Hughes Technical Center (FAA)	100
A.2.2	MITRE	103
A.2.3	NASA - Ames	106
Appendix B:	**Guidance Summary**	**109**
B.1	**Surface Map Checklist**	**109**
B.2	**Referenced CFRs**	**122**

LIST OF TABLES AND FIGURES

Table 1. Elements of each consideration. 2
Table A-1. Industry vendors. 68
Table A-2. Research organizations. 69
Table A-3. Summary of display element depiction: Vendor displays. 70
Table A-4. Summary of display element depiction: Research displays. 73
Table A-5. Summary of indicators: Vendor displays. 75
Table A-6. Summary of indicators: Research displays. 75
Table A-7. Summary of functionality: Vendor displays. 76
Table A-8. Summary of functionality: Research displays. 77
Table A-9. ACSS display elements. 78
Table A-10. Diehl Avionik display elements. 80
Table A-11. Garmin AT AT2000 display elements. 83
Table A-12. Garmin AT AT2000 indicators. 84
Table A-13. Honeywell display elements. 86
Table A-14. JeppView FliteDeck display elements. 88
Table A-15. Jeppesen TPA display elements. 90
Table A-16. Jeppesen TPA indicators. 90
Table A-17. Rockwell Collins in cockpit display elements. 93
Table A-18. Rockwell Collins indicators. 93
Table A-19. Rockwell Collins PC-based surface map display elements. 95
Table A-20. Rockwell Collins PC-based display functionality. 96
Table A-21. Smiths Avionics display elements. 98
Table A-22. FAA Tech Center display elements. 101
Table A-23. FAA Tech Center: Indicators. 101
Table A-24. MITRE surface moving map elements. 104
Table A-25. MITRE indicators. 104
Table A-26. T-NASA display elements. 107
Table B-1. 14 CFR § 23. 122
Table B-2. 14 CFR § 25. 124
Table B-3. 14 CFR § 27. 126
Table B-4. 14 CFR § 29. 127

Figure A-1. Control mode for ACSS map display. 78
Figure A-2. Diehl Avionik System display. Photo courtesy of Diehl Avionik. 80
Figure A-3. MX20 display. Photo courtesy of Garmin AT. 82
Figure A-4. MX20 controls. Photo courtesy of Garmin AT. 82
Figure A-5. Garmin AT AT2000 control layout. 84
Figure A-6. Garmin AT control functionality. 85
Figure A-7. Honeywell controls. 86
Figure A-8. (a) JeppView airport diagram, (b) Garmin AT MX-20 chart view with ownship position superimposed on a JeppView instrument approach chart. 88
Figure A-9. Jeppesen TPA System. Photo courtesy of Jeppesen. 89
Figure A-10. Rockwell Collins display. Photo courtesy of Rockwell Collins. 92
Figure A-11. Rockwell Collins control panel. Photo courtesy of Rockwell Collins. 94
Figure A-12. Rockwell Collins PC-based surface moving map. Photo courtesy of Rockwell Collins. 95
Figure A-13. Smiths Aerospace Flight Management Computer System taxi plan. Photo courtesy of Smiths Aerospace. 98
Figure A-14. FAA Tech Center surface map. Photo courtesy of William J. Hughes Technical Center..... 100
Figure A-15. FAA Tech Center: controls and functionality. 102
Figure A-16. MITRE surface display. Photo courtesy of MITRE. 103
Figure A-17. T-NASA surface display. Photo courtesy of NASA-Ames. 106
Figure A-18. T-NASA display location for two crew simulations. Photo courtesy of NASA-Ames......... 108

EXECUTIVE SUMMARY

This document is intended to facilitate the identification and resolution of flight-deck human-factors issues associated with presenting an airport surface moving map display that depicts ownship position.

The Federal Aviation Administration (FAA) specifies requirements for the approval of the surface moving map function in Technical Standard Order (TSO)-C165, *Electronic Map Display Equipment for Graphical Depiction of Aircraft Position*. TSO-C165 references RTCA DO-257A, *Minimum Operational Performance Standards for the Depiction of Navigational Information on Electronic Maps*. The guidance in this document supplements that specified in TSO-C165 and RTCA DO-257A with FAA regulations, industry documents, and user interface design principles that describe good practices relevant to the design of the surface moving map displays. This document applies to *all* surface moving map displays regardless of the display platform.

The topics in this document cover general user interface issues, the design of individual surface-moving-map display elements (e.g., the depiction of ownship, runways, taxiways, etc.), the presentation of traffic, and the usability of functions. Note that this document is not regulatory in itself; compliance with the guidance in the report is not mandated. If the display is intended to be used for functions beyond those addressed in this document, such as runway incursion alerting, land and hold short operations, or surface navigation guidance, additional guidance may apply.

Two appendices are provided in this document. Appendix A is an industry overview, updated in August, 2003, that presents a snapshot of the efforts by manufacturers and research organizations to develop a moving map display with ownship position. For each display, the elements depicted, how they are depicted, and the functionality provided is listed. Appendix B presents a summary of the requirements and recommendations listed in this document. This summary can be used by manufacturers and aviation authorities when conducting human factors evaluations of surface moving map displays.

ACRONYMS

ADS-B	Automatic Dependent Surveillance – Broadcast
ARP	Aerospace Recommended Practice
CDTI	Cockpit Display of Traffic Information
CFR	Code of Federal Regulations
EFB	Electronic Flight Bag
FAA	Federal Aviation Administration
FMS	Flight Management System
FBO	Fixed Base Operator
FIS-B	Flight Information Service – Broadcast
GNSS	Global Navigation Satellite System
JAA	Joint Aviation Authorities
NACO	National Aeronautical Charting Office
NGS	National Geodetic Survey
SAE	Society of Automotive Engineers
TCAS	Traffic Alert and Collision Avoidance System
TIS-B	Traffic Information Service – Broadcast
TSO	Technical Standard Order

1 INTRODUCTION

Many manufacturers are developing moving map displays that depict ownship position on the airport surface. Examples of surface moving map displays are presented in an industry overview (Appendix A). The intended function of each display is specified by the avionics manufacturer during the certification process.

The Federal Aviation Administration (FAA) specifies requirements for the approval of the surface moving map function in Technical Standard Order (TSO)-C165, *Electronic Map Display Equipment for Graphical Depiction of Aircraft Position*. TSO-C165 references RTCA DO-257A. This document provides additional guidance for the design and evaluation of surface moving map displays. The guidance in this document applies to *all* surface moving maps regardless of the display platform. If the surface moving map is presented on an Electronic Flight Bag (EFB), see also FAA Advisory Circular (AC) 120-76A, *Guidelines for the Certification, Airworthiness, and Operational Approval of Electronic Flight Bag Computing Devices*, and *Human Factors Considerations in the Design of Electronic Flight Bags (EFBs)* by Chandra, Yeh, Riley, and Mangold (2003).

Topics in this document address general user interface issues, the design of surface moving map display elements (e.g., the depiction of ownship, runways, taxiways, etc.), the presentation of traffic, and the usability of functions. Note that this document is not regulatory in itself; compliance with the guidance in the report is not mandated. If the display is intended to be used for functions beyond those addressed in this document, such as runway incursion alerting, land and hold short operations, or surface navigation guidance, additional guidance may apply.

1.1 Background

The FAA has requested human factors guidance to support the new moving map TSO for the wide range of implementations and functions. This document was developed to meet that need. Implementations of the surface moving map function vary in the level of detail with which the airport surface is depicted and in the functions that are available. The simplest surface moving map may consist of ownship position superimposed on a geo-referenced electronic airport diagram. The airport attributes depicted will be identical to what is presented on a paper chart. This map may be non-interactive; functionality, if any, will be limited, e.g., zooming may be available but more complicated features such as decluttering may not be available.

More complex surface moving map displays are constructed from a database that contains positional data describing the location of airport attributes. The database information is collected through a detailed survey that maps the location of these attributes. Surface moving maps constructed from a database vary in the detail with which the airport is depicted and in the functionality available. Common display elements depicted include ownship position, runways, runway labels, taxiways, taxiway labels, non-movement areas, and buildings. Common functions include zooming, displaying traffic, and decluttering. More advanced features such as surface navigation guidance with the display of a taxi route and autozoom are available for some displays.

As with any new display introduced into the flight deck environment, use of the surface moving map could have negative consequences if it is not implemented appropriately. Poor usability could result in increased workload and increased head-down time. For example, there is some evidence to suggest that a surface moving map may alter pilots' attention to the out-the-window scene (Battiste, et al., 1996a, 1996b; FAA, 2001; Hooey, Foyle, and Andre, 2000). In particular, the presentation of ownship position on a surface map display may be compelling to the degree that the pilot fails to look out-the-window to verify position and moves the aircraft based on inaccurate information depicted on the display. While data on use of surface moving maps is limited, pilot reliance on flight management systems (FMS) has shown that as pilots have become more dependent on FMS for navigation, errors in the FMS data are "… less likely to be noticed by the crew, … particularly in a rapidly changing, dynamic environment" (Coyle, 1997). Thus, pilots' tendency to overrely on advanced functions need to be considered in the design and evaluation of surface moving maps. This document intends to address these types of issues by providing guidance in the following sections.

1.2 How to Use this Document

This document provides human factors guidance for the design and evaluation of moving map displays that depict ownship position on the airport surface. The guidance provided here is a collection of FAA regulations and industry documents containing design guidelines and user interface principles. The considerations here address the following topics:

- General: These considerations discuss general user interface design principles, as they are relevant to the surface moving map function.
- Surface-Moving-Map Display Elements: This section provides design guidance for those elements that have commonly been depicted on surface moving maps that depict ownship position. The section begins with a review of database and accuracy requirements for the depiction of display elements. Considerations related to the representation of individual display elements follow.
- Traffic Display: This section provides preliminary guidance on the depiction of traffic aircraft and vehicles. While the presentation of traffic information is outside the scope of RTCA DO-257A, the depiction of traffic has been included on surface moving map displays. This guidance is provided in lieu of any published material by aviation authorities.
- Functionality: This section provides considerations to improve the usability of functions available on surface moving map displays.

In each section, considerations may be provided in one or more boxed summary statements. Each statement is preceded by a descriptive label, which identifies the type of information provided. The information in a consideration may be one of five types: FAA Policy and Guidance, Recommendations, Suggestions, or Design Tradeoffs. Table 1 describes the elements in each consideration.

FAA Policy and Guidance

- FAA related policy are shaded and boxed with a double line. The guidance here includes material from FAA regulations, TSOs, and documents invoked or referenced in FAA documents (e.g., RTCA DO-257A which is invoked by TSO C-165).
- 14 CFR §§ 23, 25, 27, and 29, which address airworthiness standards for aircraft and rotorcraft, are referenced where appropriate. The exact wording for each regulation referenced in this document can be found in Appendix B: Guidance Summary, Section B.2.
- Compliance with FAA regulations is mandatory. Compliance with the requirements in TSO C-165 is mandatory for those who choose to obtain that TSO. The use of the term "shall" in this document is only used to indicate items that are required as a means of compliance with RTCA DO-257A, as invoked by TSO C-165. Compliance with "should" statements is not mandatory.

Recommendations

- Recommendations are boxed within a bold outline. Recommendations include guidelines that affect information interpretation and/or the user's ability to access the necessary information. Compliance with this guidance is strongly encouraged, but not mandatory. Compliance with recommendations produces a better system, but increased cost or lack of feasibility may deter some designers or operators from implementing them.

Suggestions

- Suggestions are boxed within a thin outline. These statements identify good practices that have been considered in industry, but may not be appropriate in all situations.

Design Tradeoffs

Design Tradeoffs are boxed within a dotted line. These statements identify design tradeoffs and issues to be considered during design and evaluation.

Table 1. Elements of each consideration.

More detail on the guidance is provided in three sections following the boxed summary statements: Problem Statement, Examples, and Evaluation Questions. The *Problem Statement* provides a description of the problem addressed by the summary statements, and the potential impact if the problem is not addressed. *Examples* contain illustrations the potential problem and list possible solutions. Options that have been implemented by industry are presented here but may not be appropriate for all displays. *Evaluation Questions* lists questions that an aviation authority representative (e.g., FAA) could use when evaluating the system. An evaluation questions is provided for each requirement and recommendation. Note that these questions only identify areas for evaluation; they do not provide details on performance metrics.

A checklist, summarizing the requirements and recommendations, is provided in Appendix B: Guidance Summary. This summary can be used by manufacturers and regulators when conducting human factors evaluations of surface moving map displays during a bench test evaluation.

2 GENERAL

These considerations discuss general user interface design principles, as they are relevant to the surface moving map function.

2.1 Use of Color

FAA Policy and Guidance

- The accepted practice for the use of red and amber is consistent with 14 CFRs 23.1322, 25.1322, 27.1322, and 29.1322 as follows: [14 CFR §§ 23.1322, 25.1322, 27.1322, 29.1322; TSO C-165/RTCA DO-257A, 2.1.6; Chandra, et al. (2003), 2.4.8]

    (a) Red shall be used only for indicating a hazard that may require immediate corrective action.

    (b) Amber shall be used only for indicating the possible need for future corrective action.

    (c) Any other color may be used for aspects not described in items a-b of this section, providing the color differs sufficiently from the colors prescribed in these items to avoid possible confusion.

    NOTES: [TSO C-165/RTCA DO-257A, 2.1.6]

    1. Requirements a & b are intended to preclude the excessive use of amber and red on the surface moving map. They are not meant to inhibit the use of red and amber for the coding of surface signs, lights, and markings.

    2. These requirements are not intended to supersede system specific requirements in other avionics documents invoked by the FAA (e.g., TSO-C151b (TAWS), TSO-C119b (TCAS), AC 20-131A (TCAS II)).

    3. For Flight Information Service (FIS) overlays, the color guidelines of RTCA SC-195 apply. RTCA DO-267 is being updated by SC-195 including guidelines on the use of color.

- Color-coded information should be accompanied by another distinguishing characteristic such as shape, location, or text. [AC 23.1311-1A; TSO C-165/RTCA DO-257A, 2.1.6]

- No more than six colors should be used for color-coding on the map display. [TSO C-165/RTCA DO-257A, 2.1.6; SAE ARP 4032; Chandra, et al. (2003), 2.4.3]

- The colors available from a symbol generator/display unit combination should be carefully selected on the basis of their chrominance separation. Research studies indicate that regions of relatively high color confusion exist between red and magenta, magenta and purple, cyan and green, and yellow and orange (amber). Colors should track with brightness so that chrominance and relative chrominance separation are maintained as much as possible over day/night operation. Requiring the flightcrew to discriminate between shades of the same color for symbol meaning in one display is not recommended. [AC 25-11, 5.a(5); Chandra, et al. (2003), 2.4.3]

    NOTE: The Airborne Multipurpose Electronic Displays TSO (C113) references SAE ARP 1068B, "Flight Deck Instrumentation, Display Criteria and Associated Controls for Transport Aircraft," which provides color discriminability values.

Recommendation(s)

- Colors on the display should be discriminable by the typical user under the variety of lighting conditions expected in a flight deck from a nominal reference design eye point. [Chandra, et al. (2003). 2.4.3]

- Each color used in a color-coding scheme should be associated with only one meaning. [Chandra, et al. (2003), 2.4.3]

- Color-coding schemes should not conflict with flight deck standards for that particular aircraft. [Chandra, et al. (2003), 2.4.3]

Recommendation(s) (continued)

> - Pure blue should not be used for small symbols, text, fine lines, or as a background color. Blue is a short wavelength color. On a display containing several colors, when blue and other short wavelength colors are in focus, all other colors at long wavelengths are out of focus, and vice versa. [Cardosi and Hannon, 1999; Chandra, et al. (2003), 2.4.3]

Suggestion(s)

> - Large areas filled with saturated colors (e.g., rectangles filled with pure red or blue) may cause eyestrain and/or afterimages and should be avoided. [Chandra, et al. (2003), 2.4.3]
> - Pure colors should not be used when the contrast ratio between the color and its surround is low (e.g., blue elements on a black background). [Cardosi and Hannon, 1999]
> - Saturated red and blue should not be presented close together to avoid a false perception of depth. [Cardosi and Hannon, 1999]
> - Upon request, a legend describing the meanings associated with different colors should be displayed. Access to a legend of colors is especially important if the colors are user customizable. [Chandra, et al. (2003), 2.4.3]

Design Tradeoff(s)

> The display brightness setting may affect color discriminability. [Chandra, et al. (2003), 2.4.3]
>
> While color is beneficial for segregating display elements, it may be possible that color could be perceived as an attentional filter so that users focus on display elements of one color, ignoring all others. This is particularly a concern if color is used to visually segregate the presentation of traffic on a surface map display. In the worst case, the pilot could focus inappropriately on aircraft of a specific target "color", at the cost of noticing other aircraft on the display.

Problem Statement

Colors that are not discriminable will increase pilot workload, head-down time, and task completion time. Color discrimination can be compromised by a variety of factors including lighting conditions, viewing angle, display quality and calibration, and size of the object. Additionally, color perception varies across individuals. As the eye ages, its ability to focus on red objects or differentiate between blue and green is reduced. In particular, pure blue is problematic for observers over age 50.

If red is used too broadly, pilots may not quickly be able recognize situations where their actions *are* time-critical. Alternative uses of red may be allowed through other system specific FAA documents, which would supersede the guidance in this document.

Example(s)

One way to ensure that colors are used redundantly with other cues is to design the system for a monochrome display first, and then add color afterwards. (CAP 708)

Unless monitors are calibrated appropriately, the appearance of a color defined by Red-Green-Blue (RGB) or Hue-Saturation-Lightness (HSL) values will vary. In order to specify a color accurately, Commision Internationale de l'Eclairage (CIE) color coordinates are used, and monitors are calibrated to ensure that they are displaying the defined color correctly.

An afterimage is the illusory color that results after focusing on an intense, high-contrast color. For example, when one stares at a solid blue object for a period of time, upon looking away, one may see a yellow afterimage in its place.

Color has been successfully used primarily as an aid for visual search or for perceptual grouping. The use of color-coding has been found to reduce search times in densely populated displays when compared with performances obtained using size, shape, or brightness coding. Additionally, color can be used to perceptually tie together display elements which are spatially separated on the display, and vice versa. For example, runways, taxiways, and other movement areas could be colored in similar shades, but distinct from the color used to indicate non-movement areas.

Colors can be selected to enhance figure-ground segregation. A display element will be salient if it is presented in a unique color that has a high contrast with the background. Note, however, that if the same color is used to code several display elements, the display elements may not be salient, regardless of the foreground-background contrast.

Some colors have strong associations in general, such as red. Some common associations for red are danger, emergency, failure, stop, no-go, and fire/hot. Also, there are flight deck conventions for the use of red. As a result, any use of red should be considered carefully so that it does not conflict with pilot expectations in the flight deck. In order for red to retain its distinctiveness in the flight deck, it is important to limit its use to the highest priority situations, such as when pilot actions are time-critical.

Evaluation Question(s)

- Is the color red reserved for situations that may require immediate pilot action?
- Is the color amber/yellow reserved for indicating caution conditions with the possibility for future corrective action?
- Are colors used sufficiently discriminable from red or amber/yellow?
- Is color used redundantly, e.g., with shape? If the display were viewed in monochrome, would it still be possible to understand all the information being conveyed?
- Are six or fewer colors used in a color-coding scheme?
- Are all colors discriminable?
- Are all colors legible and discriminable under the variety of lighting conditions expected in the flight deck and from a nominal reference design eye point?
- Is each color in a color-coding scheme associated with only one meaning?
- If color-coding is used, are color-coding schemes designed so that they do not conflict with flight deck standards for the aircraft?
- Is the use of pure blue for small symbols, text, fine lines, or as a background color avoided?

2.2 Alerts and Reminders

FAA Policy and Guidance

- Warning information must be provided to alert the crew to unsafe system operating conditions, and to enable them to take appropriate corrective action. Systems, controls, and associated monitoring and warning means must be designed to minimize crew errors which could create additional hazards. [14 CFR §§ 23.1309(b)(3), 25.1309(c), 29.1309(c)]
- If a visual indicator is provided to indicate malfunction of an instrument, it must be effective under all probable cockpit lighting conditions. [14 CFR §§ 23.1321(e), 25.1321(e), 27.1321(d), 29.1321(g)]
- Short term flashing symbols (approximately 10 seconds or flash until acknowledge) are effective attention getters. A permanent or long term flashing symbol that is noncancellable should not be used. [AC 25-11, 5.g(1)]
- Messages should be prioritized and the message prioritization scheme should be documented and evaluated. [AC 120-76A, 10.d(1) and 10.d (2); Chandra, et al. (2003), 2.4.8]

    NOTE: Further guidance on alerts and reminders can be found in FAA standards for electronic displays in AC 25-11. International recommended guidance can be found in publications of the Joint Aviation Authorities (JAA) including AMJ 25-11 and AMJ 25.1322 on warnings and cautions. See also SAE ARP 4102/4 and the FAA DOT/FAA/CT-03/05. [Chandra, et al., 2003; 2.4.8]

Recommendation(s)

- Any use of alerts should be assessed in terms of ease of interpretation, confusion with other alerts, and for consistency with flight deck alerting philosophy.

Problem Statement

Additional hazards could occur if the warning or alert is not noticed or cannot be easily interpreted.

If messages are not prioritized, the flight crew, when presented with multiple alerts, could diagnose one and miss a more critical error.

Example(s)

One implementation of a runway occupancy alerting scheme is to highlight the edges of the occupied runway edges in red. However, since aircraft move on and off runways continuously during normal operations, the end result of the implementation is that the runway occupancy bars flash on/off. This flashing is inappropriate; it creates a condition that may be too distracting. Additionally, the use of red to signal runway occupancy may *not* be appropriate as runways at busy airports may be occupied more often than not and runway occupancy typically does not require pilot action, just pilot awareness.

If the surface moving map is presented on an installed flight deck display or EFB, messages that are time-critical may be displayed as part of the integrated warning system display. Messages specific to the EFB, however, should be displayed on the EFB.

An alert prioritization scheme from the Terrain Awareness and Warning System (TAWS) TSO-C151b is provided below as an example.

ALERT PRIORITIZATION SCHEME

Priority	Description	Alert Level [b]	Comments
1	Reactive Windshear Warning	W	
2	Sink Rate Pull-Up Warning	W	continuous
3	Excessive Closure Pull-Up Warning	W	continuous
4	RTC Terrain Warning	W	
5	V_1 Callout	I	
6	Engine Fail Callout	W	
7	FLTA Pull-Up warning	W	continuous
8	PWS Warning	W	
9	RTC Terrain Caution	C	continuous
10	Minimums	I	
11	FLTA Caution	C	7 s period
12	Too Low Terrain	C	
13	PDA ("Too Low Terrain") Caution	C	
14	Altitude Callouts	I	
15	Too Low Gear	C	
16	Too Low Flaps	C	
17	Sink Rate	C	
18	Don't Sink	C	
19	Glideslope	C	3 s period
20	PWS Caution	C	
21	Approaching Minimums	I	
22	Bank Angle	C	
23	Reactive Windshear Caution	C	
Mode 6 [a]	TCAS RA ("*Climb*", "*Descend*", etc.)	W	continuous
Mode 6 [a]	TCAS TA ("*Traffic, Traffic*")	C	Continuous

NOTE 1: These alerts can occur simultaneously with TAWS voice callout alerts.

NOTE 2: W = Warning, C = Caution, A = Advisory, I = Informational

Evaluation Question(s)

- Are warnings provided to alert the crew to unsafe operating conditions? Do the warnings enable the crew to take appropriate corrective action? Are systems, controls, and associated monitoring and warning means designed to minimize crew errors?

- Are visual indicators that indicate the malfunction of an instrument visible under all probable cockpit lighting conditions?

- Has flashing been avoided? If flashing text or symbols are presented, has the use of flashing been evaluated?

- Are messages prioritized? Is the message prioritization scheme documented? Has the message prioritization scheme been evaluated?

- Has the use of alerts been evaluated to ensure that they are easy to interpret and consistent with the flight deck alerting philosophy?

2.3 Accessibility of Controls

FAA Policy and Guidance

> - Each cockpit control must be located to provide convenient operation and to prevent confusion and inadvertent operation. [14 CFR §§ 25.777(a), 27.777(a), 29.777(a); TSO C-165/RTCA DO-257A, 2.1.5.1]
>
>   Related Policy: 14 CFR § 23.777(a) is worded slightly differently.
>
> - The use of controls should not cause inadvertent activation of adjacent controls. [TSO C-165/RTCA DO-257A, 2.1.5.1]
>
> - The controls must be located and arranged, with respect to the pilot's seats, so that there is full and unrestricted movement of each control without interference from the cockpit structure or the clothing of the minimum flight crew when any member of this flight crew, from 5'2" to 6'3" in height, is seated with the seat belt and shoulder harness fastened. [14 CFR § 25.777(c)]
>
>   Related Policy: 14 CFR §§ 23.777(b), 27.777(b), and 29.777(b) are slightly different.
>
> - Each flight, navigation, and powerplant instrument for use by any pilot must be plainly visible to him from his station with the minimum practicable deviation from his normal position and line of vision when he is looking forward along the flight path. [14 CFR §§ 25.1321(a), 29.1321(a)]
>
>   Related Policy: 14 CFR §§ 23.1321(a) and 27.1321(a) are worded slightly differently.
>
> - Controls that are normally operated by the flight crew shall be readily accessible. [TSO C-165/RTCA DO-257A, 2.1.5.2]

Problem Statement(s)

Controls that are not easily accessible could result in increased crew workload, head-down time, and inadvertent activation of other controls.

Example(s)

Angled panels allow more controls to be placed within access of the pilot than non-angled panels. A comparison of performance with varying degrees of angle for side panels (35°, 45°, 55°, and 65°) revealed that the average number of seat movements and average number of seat displacements was lowest for panels oriented at 65° (Siegel and Brown, 1958).

Sharp and Hornseth (1965) asked seated subjects to use three different types of controls (knobs, toggle switches, and pushbuttons) presented at 12 different locations on a console. The results showed that the time needed to activate a control varied as a function of the distance from the control to the subject. Activation time was at a minimum when the controls were located within 25° of the center.

Evaluation Question(s)

- Are controls located so that they are easy to reach and operate? Are controls located so that they cannot be activated accidentally?

- Are controls arranged so that the selection of one control does not result in the accidental activation of adjacent controls?

- Are controls located so that the crew can access the control without interference from the cockpit structure or clothing?

- Are controls viewable from the pilot's position and line of vision when s/he is looking forward along the flight path?

- Are flight crew controls readily accessible?

2.4 Design of controls
FAA Policy and Guidance

- The equipment must allow each flight crew member to perform their duties without unreasonable concentration or fatigue. [14 CFR § 25.771(a)]

  Related Policy: 14 CFR §§ 23.771(a), 27.771(a), and 29.771(a) are worded slightly differently.

- Each cockpit control, other than primary flight controls and controls whose function is obvious, must be plainly marked as to its function and method of operation. [14 CFR §§ 25.1555(a), 27.1555(a), 29.1555(a); Chandra, et al. (2003), 2.5.2]

  Related Policy: 14 CFR § 23.1555(a) is worded slightly differently.

- Each item of installed equipment must be labeled as to its identification, function, or operating limitations, or any applicable combination of these factors. [14 CFR §§ 23.1301(b); 25.1301(b); 27.1301(b); 29.1301(b); TSO C-165/RTCA DO-257A, 2.1.5.1]

- If a control can be used for multiple functions, the current function shall be indicated either on the display or on the control. [TSO C-165/RTCA DO-257A, 2.1.5.1]

- Line select function keys should acceptably align with adjacent text. [TSO C-165/RTCA DO-257A, 2.1.5.2]

- For each instrument, each instrument marking must be clearly visible to the appropriate crewmember. [14 CFR § 25.1543(b)]

  Related Policy: 14 CFR §§ 23.1543(b), 27.1543(b), and 29.1543(b) are worded slightly differently.

- The instrument lights must provide sufficient illumination to make each instrument, switch and other device necessary for safe operation easily readable unless sufficient illumination is available from another source. [14 CFR § 25.1381(a)(1)]

  Related Policy: 14 CFR §§ 23.1381(a)(1), 27.1381(a)(1), and 29.1381(a)(1) are worded slightly differently.

- The equipment shall be designed so that controls intended for use during flight cannot be operated in any position, combination or sequence that would result in a condition detrimental to the equipment or operation of the aircraft. [TSO C-165/RTCA DO-257A, 2.1.5.1]

- Controls shall provide feedback when operated. [TSO C-165/RTCA DO-257A, 2.1.5.1]

  NOTE: Tactile and visual cues are acceptable forms of feedback. Aural cues may also be acceptable. [TSO C-165/RTCA DO-257A, 2.1.5.1]

- Control operation should allow sequential use without unwanted multiple entries. [TSO C-165/RTCA DO-257A, 2.1.5.1]

- Manual controls used in flight shall be operable with one hand. [TSO C-165/RTCA DO-257A, 2.1.5.1]

- Activation or use of a control should not require simultaneous use of two or more controls in flight (e.g., pushing two buttons at once). [TSO C-165/RTCA DO-257A, 2.1.5.1]

- Controls should be designed for nighttime usability (e.g., illuminated). [TSO C-165/RTCA DO-257A, 2.1.5.1]

  NOTE: Control illumination may be achieved by either illuminating the control itself or providing flight deck (external) illumination. This will need to be evaluated on an installation specific basis. [TSO C-165/RTCA DO-257A, 2.1.5.1]

FAA Policy and Guidance (continued)

- Each pilot compartment must be arranged to give the pilots a sufficiently extensive, clear, and undistorted view, to enable them to safely perform any maneuvers within the operating limitations of the airplane, including taxiing takeoff, approach, and landing. [14 CFR § 25.773(a)(1)]

  Related Policy: 14 CFR §§ 23.773(a), 27.773(a)(1), and 29.773(a)(1) are worded slightly differently.

- Each pilot compartment must be free of glare and reflection that could interfere with the normal duties of the minimum flight crew. [14 CFR § 25.773(a)(2)]

  Related Policy: 14 CFR §§ 23.773(a)(2), 27.773(a)(1), 29.773(a)(2) are worded slightly differently.

- Letter keys on a keypad should be arranged alphabetically or in a QWERTY format. [TSO C-165/RTCA DO-257A, 2.1.5.3]

- If a separate numeric keypad is used, the keys should be arranged in order in a row or in a 3X3 matrix with the zero at the bottom. [TSO C-165/RTCA DO-257A, 2.1.5.3]

- If non-alphanumeric special characters or functions are used, dedicated keys should be provided (e.g., space, slash (/), change sign key (+/-), "clear" and "delete," etc.). [TSO C-165/RTCA DO-257A, 2.1.5.3]

- Where knob rotation is used to control cursor movement, sequence through lists, or cause quantitative changes, the results of such rotation should be consistent with established behavior stereotypes (Reference Sanders & McCormick, 1987) as follows: [TSO C-165/RTCA DO-257A, 2.1.5.4]

    a) For X-Y cursor control (e.g., moving a pointer across the surface of the map):
    - Knob below or to the right of the display area: clockwise movement of the knob moves the cursor up or to the right.
    - Knob above the display area: clockwise rotation of knob moves cursor up or to the left.
    - Knob to left of display area: clockwise rotation of knob moves cursor down or to the right.

    b) For quantitative displays, clockwise rotation increases values.

    c) For alphabet character selection or alphabetized lists, clockwise rotation sequences forward.

- Concentric knob assemblies should be limited to no more than two knobs per assembly. [TSO C-165/RTCA DO-257A, 2.1.5.4]

Recommendation(s)

- The shape of the control should be unique and, where possible, meaningful so it can be identified directly with the function.

- Soft function keys that are inactive should either not be labeled, or use some kind of display convention to indicate that the function is not available. [Chandra, et al. (2003), 2.5.2]

- Soft function keys are typically used as multi-function keys to select one of several available functions. When the same type of function is accessed from different points in the software, the common function should appear on the same physical function key whenever possible (e.g., top right). [Chandra, et al. (2003), 2.5.2]

NOTE: *Soft function keys* are physical push buttons whose functions can be reassigned via software. Soft function keys save space by allowing keys to be re-mapped accordingly.

Suggestion(s)

- Common and acceptable means of reducing the likelihood of inadvertent operation through key design include the following: [TSO C-165/RTCA DO-257A, 2.1.5.1]
  a) A minimum edge-to-edge spacing between buttons of 1/4 inch. (Keys should not be spaced so that sequential use is awkward or error prone.)
  b) Placing fences between closely spaced adjacent controls.
  c) Concave upper surface of keys to reduce slippage.
  d) Size of control surface sufficient to provide for accurate selection.
- Concentric knobs are typically designed so that the one closest to the face of the panel changes cursor position, selects information category, operating/display mode, or large value changes. The inner/smaller knob is used to select among the information contents, sub categories of the position selected with the outer knob, or fine value changes.
- If soft function keys are presented on a touch screen display, the active area may need to be larger than they would need to be in a stable environment for use in a turbulent en route environment.

Design Tradeoff(s)

If the display glass is set inside a bezel, the depth of the bezel frame can introduce a perceived misalignment between soft key labels and physical function keys when viewing the display off-angle (i.e., parallax errors). [Chandra, et al. (2003), 2.5.2]

Problem Statement(s)

Controls that are not labeled will increase the time to complete a task, the potential for errors, and pilot workload. Expectations about the control's action play an important role in ease of use. For this reason, as well as to ensure compliance with 14 CFR §§ 23.1555(a), 25.1555(a), 27.1555(a), and 29.1555(a), all controls should be labeled with a meaningful label so that the user knows what to expect when using them.

Poorly designed controls will result in input errors. Pilots will need to verify their entries more carefully and spend time correcting incorrect entries.

The designer must take into account several parameters of the control design so that physical controls are not too easy or too difficult to activate. These parameters include the dimensions of the control surface, spacing between controls, force required to activate controls, and displacements for displacement controls such as toggle switch.

Example(s)

If a function is mapped to different keys at different times or in different states, errors may occur because the user expects the function to be assigned to the same key all the time.

Multiple entries are sometimes registered by the hardware when a user holds down a button longer than usual, or when the user's finger hits the button twice inadvertently (e.g., in turbulence). A good rule of thumb is to discard multiple entries that occur within 300 milliseconds of each other, which is a typical length for the time between fast, discrete, intentional movements.

Backlighting can be applied to buttons and bezel knobs so that the controls are visible at night and in low illuminations (e.g., see Garmin AT MX-20).

Two ways to arrange keys in a numeric keypad are shown below. [TSO C-165/RTCA DO-257A, 2.1.5.3]

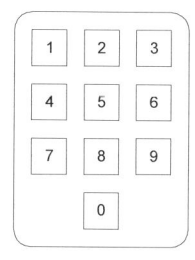

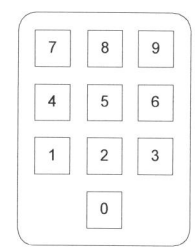

Telephone Style   **Calculator Style**

Evaluation Question(s)

- Does the equipment allow the flight crew to perform their duties without unreasonable concentration or fatigue?
- Are all controls labeled?
- If a control can be used for multiple functions, is the current function indicated on the display or the control?
- Do line select function keys line up acceptably with adjacent text?
- Are markings for each control clearly visible?
- Are controls and labels readable in all flight deck lighting conditions? Are controls designed for nighttime usability? Can control labels be read in night lighting conditions?
- Are controls designed so that those used during flight cannot be operated in any position, combination, or sequence that would result in a condition detrimental to the equipment or aircraft operations?
- Do all controls provide feedback when operated?
- Can controls be operated sequentially without creating unwanted multiple entries?
- Are manual controls that are used in flight operable with one hand?
- Can all controls be activated without the simultaneous use of two or more controls in flight?
- Do pilots have a clear and undistorted view to safely perform maneuvers?
- Is each pilot compartment free of glare and reflection that could interfere with the normal duties of the minimum flight crew?
- Are letter keys on a keypad arranged alphabetically or in a QWERTY format?
- If a separate numeric keypad is used, are the keys arranged in order in a row or in a 3x3 matrix with zero at the bottom?
- Are dedicated keys for non-alphanumeric special characters or functions provided?
- Is the direction with which a control is moved congruent with the description of that control and the resulting action?
- If concentric knobs are used, are there two or less knobs for each control?
- Is the shape of the control unique and meaningful so that it can be identified directly with the function?
- Are soft function keys that are inactive drawn so that it is clear that the function is not available?
- Are all soft function keys labeled? For soft function labels, does the label always reflect the current intended function?
- If soft function keys are used, do common functions appear on the same physical function key?

2.5 Design of Labels

Labels on a surface moving map include not only identification of controls but also identification of display elements, e.g., runway and taxiway identifiers, building names, and data tags for traffic.

FAA Policy and Guidance

- Labels shall be used to identify fixes, other symbols, and other information, depicted on the display, where appropriate. [TSO C-165/RTCA DO-257A, 2.2.2]
- The spatial relationships between labels and the objects that they reference should be clear, logical, and, where possible, consistent. [TSO C-165/RTCA DO-257A, 2.2.2; Chandra, et al. (2003), 2.5.2]
- Alphanumeric fonts should be simple and without extraneous details (e.g., sans serif) to facilitate readability. [TSO C-165/RTCA DO-257A, 2.2.2]
- Fix labels shall be oriented to facilitate readability. [TSO C-165/RTCA DO-257A, 2.2.2]
  NOTES: [TSO C-165/RTCA DO-257A, 2.2.2]
  1. One method of compliance is to continuously maintain an upright orientation.
  2. This requirement does not apply to RAC data because they may not be able to meet this requirement due to the fundamental nature of the data. It does apply to vector data superimposed onto a Raster Chart.
- Label terminology and abbreviations used for describing control functions and identifying display controls should be consistent with ICAO 8400/5 (a subset of which is included in RTCA DO-257A, Appendix A). [TSO C-165/RTCA DO-257A, 2.2.2]
- All labels shall be readable at a viewing distance of 30 inches under the full range of normally expected flight deck illumination conditions (Reference MIL STD 1472D and SAE AIR1093). [TSO C-165/RTCA DO-257A, 2.2.2]
  NOTE: The size of numbers and letters required to achieve acceptable readability may depend on the display technology used. [TSO C-165/RTCA DO-257A, 2.2.2]

Recommendation(s)

- Soft function key labels should be drawn in a reserved space outside of the main content area. [Chandra, et al. (2003), 2.5.2]
- Labels used to identify the action associated with a soft function key should be clear to the user and brief. [Chandra, et al. (2003), 2.5.2]
- Lines should be used to connect soft labels to the control buttons they identify to minimize parallax issues.

Problem Statement

Labels that are unreadable or poorly designed could increase the time to complete a task, the potential for errors, and pilot workload.

Example(s)

One way to facilitate readability is to draw labels so that they remain upright continuously.

The Garmin AT MX-20 (See Figure A-4) and Jeppesen Taxi Position Awareness (TPA) display (See Figure A-9) units have lines carved into the bezel that extend from the control button to the corresponding soft label. These lines minimize the potential for parallax and help the user identify which buttons correspond to which label.

Evaluation Question(s)

- Are labels used to identify fixes, symbols, and other information depicted?
- Is the spatial relationship between the labels and the objects they reference clear, logical, and consistent?

- Are simple alphanumeric fonts used?
- Are fix labels oriented to facilitate readability, e.g., do labels maintain an upright orientation?
- Are label terminology and abbreviations used consistent with ICAO 8400/5 (a subset of which is in RTCA DO-257A, Appendix A)?
- Are labels readable at a viewing distance of 30 inches and under the full range of normally expected flight deck illumination conditions?
- Are soft function keys labels drawn in a reserved space outside of the main content area?
- Are soft function key labels clear and brief?
- Are soft labels connected, e.g., with lines, to the control buttons they identify?

2.6 Control layout

FAA Policy and Guidance

> - Controls should be organized in logical groups according to function and frequency of use. [TSO C-165/RTCA DO-257A, 2.1.5.2]
> - Controls most often used together should be located together. [TSO C-165/RTCA DO-257A, 2.1.5.2]
> - Controls used most frequently should be the most accessible. [TSO C-165/RTCA DO-257A, 2.1.5.2]
> - Dedicated controls should be used for frequently used functions. [TSO C-165/RTCA DO-257A, 2.1.5.2]

Suggestion(s)

- Controls should be organized so that they are collocated with the displays.

Problem Statement(s)

A poor control layout will increase the training time required for the user to become familiar with the system, increase task completion time, and increase the potential for errors.

Example(s)

Controls for related actions may be grouped together, as shown below.

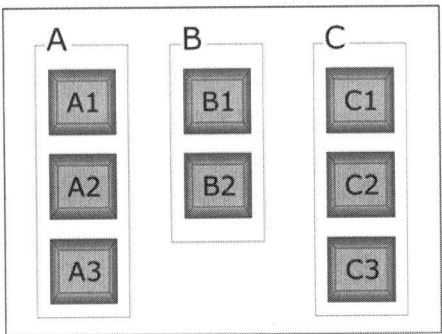

In the figure, **A**, **B**, and **C** represent three different functions. Actions for each function are provided using control buttons (**A1**, **A2**, **A3**, etc.). These buttons are organized so that related actions are next to each other.

Evaluation Question(s)

- Are controls grouped according to function? Are controls grouped according to frequency of use?
- Are controls that are used together located together?
- Are the controls used most frequently the most accessible?
- Are dedicated controls provided for frequently used functions?

2.7 Presentation of Text Information

Recommendation(s)

- The typeface size should be appropriate for the viewing conditions (e.g., viewing distance and lighting conditions) and the criticality of the text. [Chandra, et al. (2003), 2.4.11]
- Text should be spaced appropriately for ease of reading. [Chandra, et al. (2003), 2.4.12]
- A highly legible typeface enables the user to quickly and accurately identify each character. The FAA *Human Factors Design Standard for Acquisition of Commercial-off-the-shelf Subsystems, Non-Developmental Items, and Developmental Systems* (DOT/FAA/CT-03/05) recommends the following: [Chandra, et al. (2003), 2.4.10]
    (a) Upper case text should be used sparingly. Upper case text is appropriate for single words, but should be avoided for continuous text. (HFDS 8.2.5.8.2)
    (b) For continuous text (e.g., sentences and paragraphs), use mixed upper and lower case characters. (HFDS 8.2.5.8.4)
    (c) Use serif fonts for continuous text if the resolution is high enough not to distort the serifs (small cross strokes at the end of the main stroke of the letter). (HFDS 8.2.5.7.5)
    (d) Sans serif fonts should be used for small text and low resolution displays. (HFDS 8.2.5.7.6)
    (e) For optimum legibility, character contrast should be between 6:1 and 10:1. Lower contrasts may diminish legibility, and higher contrasts may case visual discomfort (HFDS 8.2.5.6.12)
    (f) Characters stroke width should be 10 to 12% of character height. (HFDS 8.2.5.6.14)
- The FAA *Human Factors Design Standard for Acquisition of Commercial-off-the-shelf Subsystems, Non-Developmental Items, and Developmental Systems* (DOT/FAA/CT-03/05) provides the following recommendations regarding the typeface size and width: [Chandra, et al. (2003), 2.4.11]
    - The minimum character height should be 16 minutes of arc (5 millirad). For practical purposes, this requires a minimum typeface height of 1/200 of the viewing distance. (DOT/FAA/CT-03/05, 8.2.5.6.6)
    - The preferred character height is 20 to 22 minutes of arc (approximately 6 millirad). For practical purposes, this translates into a typeface height of 1/167 of the viewing distance. (DOT/FAA/CT-03/05, 8.2.5.6.5)
    - The ratio of character height to width should be:
        a) At least 1:0.7 to 1:0.9 for equally spaced characters and when lines of 80 or fewer characters are used.
        b) At least 1:0.5 if more than 80 characters per line are used.
        c) As much as 1:1 for inherently wide characters such as "M" and "W" when proportionally spaced characters are used.
    If these guidelines are not met, there should be a sound basis for deviation.
- In order to make text easily readable, the FAA *Human Factors Design Standard for Acquisition of Commercial-off-the-shelf Subsystems, Non-Developmental Items, and Developmental Systems* recommends the following:
    - Use a horizontal spacing between characters of at least 10 percent of character height. (DOT/FAA/CT-03/05, 8.2.5.6.1)
    - Use spacing between words of at least one character when using equally spaced characters or the width of the capital letter "N" for proportionally spaced characters. (DOT/FAA/CT-03/05 8.2.5.6.2)
    - Use a vertical spacing between lines of at least two stroke widths or 15 percent of character height, whichever is larger. Vertical spacing begins at the bottom of character descenders (that part which descends below the text line as seen in the lower-case letter "y") and ends at the top of accent marks on upper case characters. (DOT/FAA/CT-03/05 8.2.5.6.3)

Problem Statement

Display elements that are not legible or easily interpreted may cause pilot distraction and increase pilot workload. The pilot may misread information or not be able to read this information at all.

If the font size is too small, users may misread the text or incur additional workload to adjust the display to make the text legible, e.g., by zooming the display in or out. Note that in low illuminations, a larger font size may be necessary to facilitate readability.

Example(s)

Legibility of characters and symbols are affected by size, shape, stroke width, font type, map range, viewing distance, and display location. Italicized typefaces and narrow characters take longer to read and are more likely to be misread than plain text.

Sans serif fonts are more legible than serif fonts when viewed on a display. Characters printed in a san serif font do not show the small horizontal strokes at the top and bottom, e.g., h and p, whereas serif fonts do, e.g., h and q.

Legibility is affected by the similarity of characters, e.g., "I (the letter)" with "1 (the number)". Other examples of confusable characters include: P/R, B/D/E, G/O/C, O (the letter)/ 0 (the number), Z/2.

Evaluation Question(s)

- Is the typeface size easily legible under normal viewing conditions? Is the typeface size adequate for emergency checklists and other important text that may be used under low-visibility conditions?
- Is text spaced appropriately for ease of reading?
- Are individual characters easily recognized for each typeface that is used? That is:
  - Is upper case text used sparingly?
  - Are mixed upper and lower case characters used for sentences and paragraphs?
  - For high resolution displays, are serif fonts used? For low resolution displays, are sans serif fonts used?
  - Is character contrast between 6:1 and 10:1?
  - Is character stroke width between 10 to 12% of the character height?
- Is the minimum character height at least 16 minutes of arc (5 milirad)? Is the ratio of character height to width consistent with the following:
  at least 1:0.7 to 1:0.9 for equally spaced characters when lines of 80 or fewer are used?
  at least 1:0.5 if more than 80 characters per line are used?
  as much as 1:1 for inherently wide characters when proportionally spaced characters are used?
- Is the horizontal spacing between characters at least 10% of the character height?
- If equally spaced characters are used, is the spacing between words at least one character in width? If proportionally spaced characters are used, is the space between characters the width of the capital letter "N"?
- Is the vertical spacing between lines at least two stroke widths or 15% of character height, whichever is larger?

2.8 Symbols

FAA Policy and Guidance

> - All symbols shall be depicted in an upright orientation except for those designed to reflect a particular compass orientation. [TSO C-165/RTCA DO-257A, 2.2.1.1]
>
>   NOTE: This requirement does not apply to RAC data because it may not be able to meet this requirement due to the fundamental nature of that data. It does apply to vector data superimposed onto a Raster Chart. [TSO C-165/RTCA DO-257A, 2.2.1.1]

Recommendation(s)

> - Symbols should be distinguishable based on their shape alone, without relying upon secondary cues such as color and text labels. [Chandra, et al. (2003), 2.4.13]
> - Symbols should be designed so that they are discriminable when presented on the minimum expected display resolution when viewed from the maximal intended viewing distance. [Chandra, et al. (2003), 2.4.13]
>
>   NOTE: SAE ARP 4102/7 on Electronic Display Symbology for EADI/PFD gives minimum symbol visual angles as 6 millirad for primary data, and 4 millirad for secondary and descriptive data. [Chandra, et al. (2003), 2.4.13]

Suggestion(s)

> - Where possible, shapes of non-text display elements should be consistent with paper symbol equivalents. [Chandra, et al. (2003), 2.4.13]
> - In order to assess legibility of symbols, the following factors should be considered:
>
>   (a) Similarity to other symbols and graphics: A symbol is distinctive if it is easy to discriminate from other symbols, even if it differs from other symbols by only one feature. The smallest visual feature of the symbol that distinguishes it from other symbols and graphics needs to be drawn so that it can be seen easily.
>
>   (b) Symbol size: The minimum size at which a symbol is presented must preserve the key features of the symbol. Detail in a symbol will be more difficult to distinguish on a small symbol than on a large symbol.
>
>   (c) The context in which the symbol will be shown: It is easier to discriminate one symbol from another than to recognize (identify) a symbol in the absence of any context. Context clues (e.g., the location of the symbol) may be used to determine what the symbol represents, if the meaning is not obvious.
>
>   (d) The importance of the information conveyed: A change in a small symbol feature, that may be easily missed, should not imply a significant difference in the operational interpretation of the symbol.
>
>   (e) The conditions under which the item might be viewed (e.g., position of the display): Some symbols may have fine detail that is difficult to see under degraded conditions.
>
>   (f) Optical qualities of the display: Display factors affecting legibility include resolution, contrast, brightness, color, and rendering techniques such as anti-aliasing.

Problem Statement

If fine symbol detail is necessary to distinguish between symbols that represent significantly different operational impacts, and that fine detail is not easily seen, the consequences could be operationally significant. Symbols presented with fine detail may be difficult to see in degraded conditions or at high map ranges (i.e., zoomed out). Even if misinterpretation does not occur, poorly designed symbols may cause confusion, and as a result, increase crew workload. [Chandra, et al. (2003), 2.4.13]

Similarity in the depiction of display elements (e.g., same fill and color) will make it more difficult for the user to discriminate between airport attributes and could lead to errors. Additionally, the user will need to look longer at the display to discriminate symbols resulting in increased heads-down time.

Example(s)

When designing new methods for presenting symbology, it is important to consider consistency with other standards, such as standards for paper charts. Pilots flying FMS-equipped aircraft may confirm information presented on the display by looking to a paper chart. Consistency in symbol-task pairings facilitates interpretation of the symbology.

Discriminability is affected by the similarity in how display elements are depicted. For example, taxiway labels, particularly characters composed primarily of linear features such as "Y", may be confused with hold short lines. Both display elements are composed of straight lines whose size subtends the width of the taxiway. Increasing the similarity between the two, e.g., drawing the two display elements in identical colors, would further hinder discriminability.

Evaluation Question(s)

- Are all symbols, except those designed to reflect a particular compass orientation, depicted in an upright orientation?
- Are symbols distinguishable by shape alone?
- Are symbols legible and interpretable at the intended screen resolution from the intended viewing distance?

2.9 Graphical Icons

Recommendation(s)

- Graphical icons should be accompanied by brief text labels if their meaning is not obvious. (See also 2.5 Design of Labels) [Chandra, et al. (2003), 2.4.4]

    NOTE: 14 CFR §§ 23.1555, 25.1555, 27.1555, and 29.1555 requires labels for all controls unless their function is obvious.

- If graphical icons are used as labels, the meaning of the icon should be obvious.
- Graphical icons should be designed carefully to minimize any necessary training, and to maximize the intuitiveness of the icon for cross-cultural use. [Chandra, et al. (2003), 2.4.4]

Suggestion

- Users should be able to access text help information to explain meaning of graphical icons in more detail than the text label alone. [Chandra, et al. (2003), 2.4.4]
- Use the same graphical icons as other flight deck systems to reduce training requirements and help to prevent error, especially in high workload phases of flight or during abnormal or emergency situations. [Chandra, et al. (2003), 2.4.4]
- Icons should be evaluated to ensure that they are understandable. Methods for testing the meaningfulness of icons are provided by International Organization for Standardization (ISO) 9186: *Graphical symbols – Test methods for judged comprehensibility and for comprehension.*

Problem Statement

If the intended meaning conveyed by the icon is not immediately clear or the form of an icon is not intuitively linked with the function represented, the usefulness of the icon will be reduced as the user will need to learn and remember the icon's meaning. This increases the training time to use the display and increases the potential for errors in high workload situations or over time, e.g., if the user does not remember an icon's meaning.

Example(s)

One method for presenting text information about an icon is to present "tool tips", a text label that appears when the mouse cursor lingers over the icon.

The International Civil Aviation Organization (ICAO) standards document, Aeronautical Charts Standards and Recommended Practices (SARPS) Annex 4, includes some requirements for the depiction of paper and electronic chart symbols.

Standard icons have also been developed for road signs (e.g., stop signs, pedestrian crossing, yield). These icons are documented in booklet from the USDOT Federal Highway Administration titled *Road User Guide for* North America (DOT FHWA-SA-99-020). The icons in the booklet are well designed in that their meanings are intuitive across cultures. Common signs are especially similar across cultures.

Evaluation Question(s)

- Are all graphical icons labeled?
- If icons are used as labels, is their meaning obvious?
- Are icons designed so that their meaning is clear to the user? If not, does the initial training adequately address icon meanings?

2.10 Configuring Display Properties

Recommendation(s)

> - If user-interface customization by the end user is supported, the end user should be provided with an easy means by which to reset all customized parameters back to their default values. [Chandra, et al. (2003), 2.4.19]
> - The current operating mode and the functionality being configured should be indicated clearly.
>
>   NOTE: Flashing the display elements or indicators that correspond to the functionality being configured may not be an appropriate means for indicating the current operating mode or the functionality being configured. Recommendations on the general use of flashing are provided in AC 25-11, and in Section 2.2 Alerts and Reminders of this document.

Problem Statement

If the current operating mode is not indicated clearly, the user could configure the display thinking s/he is in one mode but in fact be in another. This lack of mode awareness could result in additional workload, increase the risk of error, and increase the amount of time necessary to configure the display.

Example(s)

On one prototype display, when the user changes the map range, the range ring flashes to indicate that it is selected. The use of flashing in this instance is distracting and therefore inappropriate.

Evaluation Question(s)

- If user-interface customization by the end user is allowed, can all parameters be reset to their default values easily?
- Is the current operating mode and the functionality being configured clearly indicated?

2.11 Failure Conditions

FAA Policy and Guidance

> - The equipment, systems, and installations whose functioning is required by this subchapter, must be designed to ensure that they perform their intended functions under any foreseeable operating condition. [14 CFR §§ 23.1309(b)(1), 25.1309(a), 27.1309(a), 29.1309(a)]
>   Related Policy: 14 CFR § 23.1309(b)(1) is worded slightly differently.
> - The airplane systems and associated components, considered separately and in relation to other systems, must be designed so that: [14 CFR §§ 23.1309(b)(2), 25.1309(b)]
>   1) The occurrence of any failure condition which would prevent the continued safe flight and landing of the airplane is extremely improbable, and
>   2) The occurrence of any other failure conditions which would reduce the capability of the airplane or the ability of the crew to cope with adverse operating conditions is improbable.
>   Related Policy: 14 CFR §§ 27.1309(b) and 29.1309(b) are worded slightly differently.
> - Compliance with the requirements of paragraph (b) of this section must be shown by analysis, and where necessary, by appropriate ground, flight, or simulator tests. The analysis must consider -- [14 CFR §§ 23.1309(b)(4), 25.1309(d)]
>   1) Possible modes of failure, including malfunctions and damage from external sources.
>   2) The probability of multiple failures and undetected failures.
>   3) The resulting effects on the airplane and occupants, considering the stage of flight and operating conditions, and
>   4) The crew warning cues, corrective action required, and the capability of detecting faults.
> - Any probable failure of the surface moving map shall not degrade the normal operation of other equipment or systems connected to it beyond degradation due to the loss of the surface moving map itself. [TSO C-165/RTCA DO-257A, 3.1.4]
> - The failure of interfaced equipment or systems shall not degrade normal operation of the surface moving map equipment beyond degradation due to the loss of data from the interfaced equipment. [TSO C-165/RTCA DO-257A, 3.1.4]
> - If an application is fully or partially disabled, or is not visible or accessible to the user due to a failure, this loss of function should be clearly indicated to the user with a positive indicator. That is, lack of an indication is not sufficient to declare a failure condition. [AC 120-76A, Section 10.d (2), Chandra, et al. (2003), 2.4.9]

Problem Statement

If no indication of a failure is given, then the user may make decisions on information that is incorrect, incomplete, or outdated.

Example(s)

If the surface moving map display shows traffic via data link, then the user should be notified if there is a problem with the data link that precludes normal display of the data. If the link is down completely, and there are no data to display, this should be distinguished from the case where there is a blank screen because, for example, there is no traffic in the selected region. If the link is operational in a degraded mode (e.g., the data rate is half of the normal rate so that the data are refreshed less often), this should also be brought to the crew's attention. (Note: For data link services, this requirement to notify the crew of system errors should be consistent with the appropriate Minimum Operational Performance Standards for that service, available through RTCA, and with any related FAA approved guidance materials, e.g., AC 20-140 on Aircraft Data Communication Systems and RTCA DO-257.) [Chandra, et al. (2003), 2.4.9]

For instance, the surface moving map cannot "lock up" due to failure of interfaced equipment. The system should gracefully degrade its display and should remain responsive to user inputs. [TSO C-165/RTCA DO-257A, 3.1.4]

Evaluation Question(s)

- Does the surface moving map perform its intended function under all operating conditions?
- Has the surface moving map been designed so that its failure will not prevent the continued safe flight and landing of the airplane? Has the surface moving map been designed so that its failure will not reduce the capability of the airplane or the ability of the crew to cope with adverse operating conditions?
- Has the surface moving map been designed so that its failure will not degrade the operation of other equipment or systems connected to it, beyond degradation due to the loss of the surface moving map function itself?
- Has the surface moving map function been designed so that the failure of interfaced equipment or systems does not degrade the normal operation of the surface moving map beyond degradation due to the loss of data from the interfaced equipment?
- Are failures annunciated with a positive indicator?

2.12 Update Rate

FAA Policy and Guidance

- For those elements of the display that are normally in motion, any jitter, jerkiness, or ratcheting effect should neither be distracting nor objectionable. [AC 25-11, 6.e]
- Movement of map information should be smooth throughout the range of aircraft maneuvers. [TSO C-165/RTCA DO-257A, 2.2.4]
- Maximum latency of aircraft position data at the time of display update shall be one second, measured from the time the data is received by the display system. [TSO C-165/RTCA DO-257A, 2.2.4]
- When the display receives a "data not valid" or "reduced performance" (e.g., dead reckoning mode) indication from the source, this condition shall be indicated on the display within one second. [TSO C-165/RTCA DO-257A, 2.2.4]

NOTE: Caution- some Global Navigation Satellite System (GNSS) receivers compliant with TSO-C129a do not provide this indication via the data bus. These position sources may continue to output last known position after a sensor failure. This is not acceptable. [TSO C-165/RTCA DO-257A, 2.2.4]

- If aircraft positioning data are not received by the display for five seconds (i.e., data timeout), this condition shall be indicated to the flight crew. [TSO C-165/RTCA DO-257A, 2.2.4]
- If there is an active flight plan and the flight plan data are not received by the display for 30 seconds, this condition shall be indicated to the flight crew. [TSO C-165/RTCA DO-257A, 2.2.4]

NOTE: Surface moving maps are not required to have flight plan information while on the airport surface. This requirement does apply to surface moving maps that have flight plan information and also to surface moving maps that depict a taxi route. Otherwise, this requirement does not apply to surface moving maps. [TSO C-165/RTCA DO-257A, 2.2.4]

- The display shall update the displayed minimum required information set at least once per second. The minimum required information set for surface moving map displays consists of ownship and runways. [TSO C-165/RTCA DO-257A, 2.2.4]

NOTES: The following exceptions apply: [TSO C-165/RTCA DO-257A, 2.2.4]

1) While the display must be capable of operating at an update rate of once per second, it is acceptable to adjust the update rate either dynamically or at installation to match the update rate of the position source. While acceptable it is not necessary to update the display more often than once per second even if the data source is being updated at a higher rate.

2) It is acceptable for a longer delay, not exceeding five seconds, to occur at state transitions (e.g., orientation mode, range, and leg changes).

3) At larger map ranges this requirement may not be necessary since the movement of the minimum required information set may not be noticeable.

Design Tradeoff(s)

If traffic aircraft are presented by the surface map, the update rate of positional information will vary depending upon the source transmitting the information (e.g., ADS-B transmits data more frequently than TIS-B). While the presentation of traffic information is useful, the variations in update rate may hinder pilots' abilities to predict traffic intent. One method that has been used to indicate the different update rates is by varying the representation of the traffic aircraft. However, this implementation increases the number of different icons that are presented on the traffic display at any given time.

Problem Statement

If the update rate and depiction is not timely, the information on the surface moving map could differ from what the pilot sees out-the-window, leading to confusion.

Example(s)

One method for depicting smooth aircraft movement is to extrapolate aircraft position between updates. The information depicted on the surface moving map should not flicker as ownship taxies or as the user zooms in and out of the display. Similarly, traffic position should update seamlessly as the map rotates to reflect ownship's position.

Some Global Navigation Satellite System (GNSS) receivers compliant with TSO-C129a do not provide a "data not valid" or "reduced performance" indication from the source via the data bus. These position sources may continue to output last known position after a sensor failure. This is not acceptable.

Evaluation Question(s)

- Is the display free of jitter, jerkiness, or any ratcheting effects?
- Is the movement of map information smooth throughout the range of aircraft maneuvers?
- Is the maximum latency of aircraft position data when it is updated one second or less?
- Is an indication of "data not valid" or "reduced performance" presented within one second of an indication of a failure from the source?
- Is an indication presented to the crew if aircraft positional information is not received within five seconds?
- If flight plan information is shown, is an indication presented to the crew when this information is not received for 30 seconds?
- Is the minimum required information set updated at least once per second? For surface moving maps, the minimum required information set consists of ownship and runways.

2.13 Responsiveness

FAA Policy and Guidance

> - The display shall respond to operator control inputs within 500 msec. [TSO C-165/RTCA DO-257A, 2.2.4]
> - It is desirable to provide a temporary visual cue to indicate that the control operation has been accepted by the system (e.g., hour glass or message). It is recommended that the system respond within 250 msec. [TSO C-165/RTCA DO-257A, 2.2.4]

Problem Statement

Without feedback, the user may enter inputs multiple times, which increases the likelihood of error. Additionally, the user may become confused when the system finally acts on what has been processed as multiple inputs.

Example(s)

Common symbols for indicating that the system is busy include clocks, hour-glasses, or spinning dials. Progress indicators may be presented graphically, e.g., with a bar that is shaded in proportion to the degree the task has been completed, or with text, e.g., time-to-complete or percent complete.

Evaluation Question(s)

- Does the display respond to control inputs within 500 msec?
- Are progress indicators presented for tasks that take longer than 250 msec to process? Do the indicators used clearly convey progress in a useful way?

2.14 Shared Display Considerations

FAA Policy and Guidance

> - The minimum flight crew must be established so that it is sufficient for safe operation, considering the workload on individual crewmembers. [14 CFR §§ 23.1523(a), 25.1523(a), 27.1523(a), 29.1523(a)]
> - Where information on the shared display is inconsistent, the inconsistency shall be obvious or annunciated, and should not contribute to errors in information interpretation. [TSO C-165/RTCA DO-257A, 2.1.9]
> - If information, such as traffic or weather, is with the navigation information on the electronic map display, the projection, the directional orientation and the map range should be consistent among the different information sets. [TSO C-165/RTCA DO-257A, 2.1.9; Chandra, et al. (2003), 6.2.8; SAE ARP 5898, 8.3.5]
> - Symbols and colors used for one purpose in one information set should not be used for another purpose within another information set. [TSO C-165/RTCA DO-257A, 2.1.9]
> - Deselection of shared information (e.g., weather, terrain, etc.) should be possible to declutter the display or enhance readability. (See also 5.3 Decluttering) [TSO C-165/RTCA DO-257A, 2.1.9]

Problem Statement

Incompatibilities between information sources in terms of orientation, scale, symbols, and colors could make information integration difficult and increase workload and pilot head down time (concentration and fatigue) trying to reconcile the two displays.

The availability of a surface map display that provides useful and compelling information could alter pilots' visual workload for monitoring events in the outside world (e.g., traffic), in such a way that pilots neglect the out-the-window view. Although pilots' attention is guided to relevant areas of the airport surface by information depicted on the surface map display, the consequences when an aircraft in the world is not depicted on the display is an issue that must be considered in all mixed equipage environments.

Example(s)

If some information is duplicated on spatial displays in the flight deck (e.g., terrain is shown both on the moving map, and on Enhanced Proximity Ground Warning Systems (EGPWS)), pilots could be confused by the fact that it is possible for the displays to provide different information (in terms of resolution or accuracy) if the underlying databases come from different sources.

Wreggit and Marsh (1998) reported that simply adding a GPS receiver to assist in navigation resulted in head-down times to the GPS of around 10 seconds. Put in the context of a flight scenario, if two aircraft were separated by 1 nautical mile, and approaching each other on a collision course at a speed of 300 knots, the collision would occur in 12 seconds; a pilot, head-down for 10 seconds, would have only 2 seconds to detect and avoid the potential collision.

A two-button activation required to access the traffic display will divert attention from other cockpit tasks, increasing head-down time.

Hooey, Foyle and Andre (2000) simulated a failure in the presentation of surface traffic and measured pilots' responses to a near incursion when the intruding aircraft did *not* appear on the T-NASA display. The time to respond to the intruding aircraft was compared to pilots' responses when the T-NASA display was not available. The results revealed that when an aircraft in the outside scene was not depicted on the T-NASA display, pilots were slower to detect the intruder when the T-NASA display was available than when it was not. This "surprise" appearance of the intruder was also reflected by a larger deceleration rate in conditions when pilots were using the T-NASA display than when it was not available.

Evaluation Questions

- How does the workload required for completing a task with the surface moving map compare with the workload for completing the task with an alternative method? If there is an increase in the workload of completing a task with the surface moving map relative to alternative methods, is this increase acceptable?

- Are inconsistencies on shared displays obvious and/or annunciated?
- Is the directional orientation and map range consistent? If not, is this inconsistency clearly indicated to the user?
- Are symbols and colors used consistently across information sets?
- Can shared information be decluttered?

3 SURFACE-MOVING-MAP DISPLAY ELEMENTS

This section provides design guidance for those elements that have commonly been depicted on surface moving maps that depict ownship position, as noted in Appendix A: Industry Overview. The section begins by reviewing database and accuracy requirements for the depiction of display elements. Considerations related to the representation of individual display elements follow in Section 3.3.

3.1 Databases

FAA Policy and Guidance

- If the airport map database is separate from the navigation information database, the surface moving map shall provide a means to identify the database version, and/or date, and/or valid operating period. [TSO C-165/RTCA DO-257A, 2.3.5]

  NOTE: An acceptable means of compliance is to require the pilot to acknowledge an out-of-date (or "expired") database upon start-up. Alternatively, a flight crew procedural check of airport map data base validity would also be acceptable. [TSO C-165/RTCA DO-257A, 2.3.5]

- The display shall indicate if any data is not yet effective or is out of date. [TSO C-165/RTCA DO-257A, 2.2.5]

- There should be a required pilot action acknowledging an expired database. [TSO C-165/RTCA DO-257A, 2.2.5]

- WGS-84 position reference system or an equivalent earth reference model shall be used for all displayed data. (Reference RTCA DO-236A and ICAO Annex 15). [TSO C-165/RTCA DO-257A, 2.2.5]

  NOTE: It is recognized that many datums exist other than ICAO Annex 15 WGS-84 and that conversions exist between various datums. However, datums and conversions other than WGS-84 cannot be approved without determining acceptable datum equivalency to WGS-84. It is the responsibility of the approving authority to determine if an alternate datum is equivalent. [TSO C-165/RTCA DO-257A, 2.2.5]

- The process of updating aerodrome databases shall meet the standards specified in RTCA DO-200A/EUROCAE ED-76. [TSO C-165/RTCA DO-257A, 2.3.5]

  NOTES: [TSO C-165/RTCA DO-257A, 2.3.5]

  1. As a component of RTCA DO-200A/EUROCAE ED-76 compliance, manufacturers of electronic map display equipage offering an aerodrome moving map depicting ownship position must define a process enabling their customers to expeditiously report any errors they experience in their display of ownship position solutions. One means of compliance with this requirement is, for aerodromes where customers report suspected errors, a manufacturer's process will define how they will attempt to verify and correct the error(s). As part of the process, the manufacturer will also define how they expeditiously disseminate corrections for the errors back to their customers. This overall process can resemble the processes used for reporting and correcting errors customers see when using other systems such as the Terrain Awareness and Warning Systems (TAWS) and airborne navigation systems.

  2. When the manufacturer identifies an error as a potential database error and they do not produce the database in-house, the manufacturer must identify the suspect error to their database supplier as specified in RTCA DO-200A/EUROCAE ED-76.

  3. The database supplier must attempt to verify or deny a reported database error as specified in RTCA DO-200A/EUROCAE ED-76.

Suggestion(s)

> - Acceptable means for indicating that a database is out of date include: [TSO C-165/RTCA DO-257A, 2.2.5]
>   1. disabling the display of out-of-date data;
>   2. using a distinct means of identifying out-of-date data on the display (e.g., unique color, shape, special label, etc.); or
>   3. indicating to the pilot during start-up which specific data is out-of-date (e.g., a. message that says "off-route data not current" or "only on-route fixes and off-route airports are current, all other data is out of date"), and indicate in the operating manual that any out-of-date data displayed on the electronic map display must either a) be verified to be correct by the flight crew before use or b) not be used. Complex start-up messages with long lists of what is out of date are not acceptable.

Problem Statement

If data from an out-of-date database is depicted, the airport depiction may be inaccurate leading to increased workload and increased error potential.

Example(s)

One way of checking whether the surface moving map database is current is to provide the valid dates for the airport map database at start-up. Another way is for the surface moving map to automatically check the data and provide the pilot with a message indicating whether the database was approved or not for the current flight. The automated checking routines would have to be verified themselves. [Chandra, et al., 2003]

Evaluation Question(s)

- Is the database version and valid operating period accessible presented clearly?
- Does the display indicate if any data is not yet effective or is out of date?
- Is pilot action required to acknowledge out-of date databases?
- Does displayed data use the WGS-84 position reference system or an equivalent earth reference model?
- Does the process for updating the database conform to the standards specified in RTCA DO-200A/EUROCAE ED-76?

3.2 Accuracy
FAA Policy and Guidance

- All displayed symbols and graphics shall be positioned (i.e., drawn or rendered) accurately relative to one another such that placement errors are less than .013 inches on the map depiction or 1% of the shortest axis (i.e., horizontal and vertical dimension) of the map depiction, and orientation errors are less than 3° with respect to the values provided by the position and database sources. [TSO C-165/RTCA DO-257A, 2.2.1]

  NOTES: [TSO C-165/RTCA DO-257A, 2.2.1]

  1. The goal of this requirement is to ensure that the display does not contribute significantly to the total system error to assure that the intended use of the display as a positional awareness tool is not diminished.
  2. RTCA DO-236A (Minimum Aviation System Performance Standards: Required for Area Navigation) addresses error sources and error terms that make up the total system error budget.
  3. RAC displays may not meet this requirement because the production processes for aeronautical charts allow for some leeway in the placement of aeronautical symbols for chart readability purposes. Thus, measures must be taken to advise the user of these inherent positioning errors.

- The display shall provide an indication if the accuracy implied by the display is better than the level supported by the total system accuracy. [TSO C-165/RTCA DO-257A, 2.3.1]

  NOTES: [TSO C-165/RTCA DO-257A, 2.3.1]

  1. The total system accuracy includes consideration of all error sources, including the positioning accuracy, the data accuracy and resolution, display resolution and addressability, latency, etc.
  2. The accuracy implied by the display depends upon the system implementation. For example, the scale of the ownship symbol relative to the map range may imply a level of accuracy. If a system provides the ability to display a circle around the ownship symbol that indicates the system accuracy, then the circle would define the implied accuracy. The system may account for the fact that the inaccuracy is not constant: for example, the accuracy of survey data may vary. The objective is to ensure the user is aware of the performance limitations of the system.
  3. Although new airport surveys are expected to provide more accurate airport data, currently the most significant error source is expected to be the data describing the airport environment. Rather than trying to validate the accuracy of data before it is used, acceptable system performance is achieved through reporting of errors, and having a process to take corrective action or notify operators when there is an unresolved error. It is expected that pilots will report errors if they observe that the indicated position is inconsistent with the accuracy implied by the display. One intent of the indication required by this paragraph to reduce the number of false data error reports, caused because the implied accuracy is better than the actual, expected accuracy.

FAA Policy and Guidance (continued)

- The total system accuracy shall be sufficient for the intended operation, and shall not exceed 100 meters (95%). [TSO C-165/RTCA DO-257A, 3.2.3]

  NOTES: [TSO C-165/RTCA DO-257A, 3.2.3]

  1. The accuracy is sufficient for the intended function if, when the aircraft is on a runway or taxiway, the surface moving map displays the ownship on that runway or taxiway. This may mean that the symbol itself is depicted overlapping the true aircraft location, or that the aircraft's true location stays within the depicted accuracy circle.

  2. The accuracy includes the effects of how latent data manifests itself into ownship position errors at the time of display. This includes the effects of a) not compensating vehicle movement during the latency period, b) not completely compensating (e.g., partial compensation), and/or c) errors in the compensation.

  3. The accuracy includes the effects of the aircraft reference point, defined as the accuracy of the location on the aircraft used to position the ownship symbol, or used in a surveillance position report.

  4. It is understood that the most significant error source of the total system error is the available survey data for airports. The achieved accuracy of the data is primarily an operational issue. More accurate survey data will eventually become available, and errors in the existing survey data will be culled as error reports are generated and resolved.

  5. The total system accuracy requirement is consistent with the runway separation criteria for large airport.

- The inaccuracies in the depiction of ownship position should be indicated by depicting a "circle of uncertainty" around the aircraft symbol. The radius of the circle should consider feature placement standards of the originating charting agency and errors introduced by the processing steps. It is recommended that the radius indicate a 2-sigma (95%) confidence level based on a numerical analysis of the inherent errors. Accuracy is also affected by the position sensor. If a position source other than GNSS is used, the position error inherent in the position sensor system must be taken into account and a corresponding increase of the radius of the circle of uncertainty may be required. [TSO C-165/RTCA DO-257A, F.3]

- It is recommended that manufacturers include text similar to the following in the user manual and/or on a product identification screen: "Note: Discrepancies [of up to Xnm] in the placement of airport and navigational aid symbols are known to exist in the source material. This product is not intended *for navigation guidance*." [TSO C-165/RTCA DO-257A, F.3]

- The aircraft position sensor horizontal positional accuracy for runways shall be less than 36m. [TSO C-165/RTCA DO-257A, 2.3.1.1.1]

  NOTES: [TSO C-165/RTCA DO-257A, 2.3.1.1.1]

  1. Horizontal positional accuracy is defined as the difference between a sensor's measured horizontal position and it's true horizontal position.

  2. The sensor horizontal positional accuracy requirement of 36 m was derived from the 95 percent horizontal performance of GPS (Reference DOD, GPS Standard Positioning Service Performance Standard, October 2001). The horizontal positional accuracy supports the total accuracy requirement described above.

  3. There are no horizontal protection limit (HPL) requirements for the position information used for the surface moving map.

  4. An acceptable method of compliance with this requirement is to demonstrate that the system is connected to any Global Navigation Satellite System (GNSS) sensor.

FAA Policy and Guidance (continued)

- The aerodrome total database accuracy for runways shall be 43m or less. [TSO C-165/RTCA DO-257A, 2.3.1.1.1]
  NOTES: [TSO C-165/RTCA DO-257A, 2.3.1.1.1]
  1. Aerodrome total database accuracy was derived as follows: (Aerodrome total database accuracy)^2 = (database accuracy)^2 + (survey accuracy)^2 where database accuracy and survey accuracy both equal 30m.
  2. The aerodrome total database accuracy supports the total accuracy requirement.
  3. An acceptable method of compliance with this requirement is to utilize data from a vendor that states the aerodrome database meets the 43m requirement.
- The aircraft position sensor horizontal positional accuracy for taxiways shall be less than 36m. [TSO C-165/RTCA DO-257A, 2.3.1.1.2]
- The aerodrome total database accuracy for taxiways shall be 65m or less. [TSO C-165/RTCA DO-257A, 2.3.1.1.2]
  NOTES: Acceptable means of compliance include: [TSO C-165/RTCA DO-257A, 2.3.1.1.2]
  1. Utilize data from a vendor that state the aerodrome database meets the 65m requirement.
  2. Compare the taxiway data for the surface moving map in question with data surveyed to a known accuracy (e.g., RTCA DO-272/ED-99). This should be done for a sampling of airports.
  3. For airports where no known taxiway data is published and errors are noted, operators using the moving map will report database errors to the database supplier.
- If runway markings (e.g., runway centerline) are provided they should be depicted in their correct relative position. [TSO C-165/RTCA DO-257A, 2.3.2]

Recommendation(s)

- The ownship symbol should only be displayed on maps or charts that are georeferenced and to scale. [Chandra, et al. (2003), 6.2.10]
- The range of display zoom levels should be compatible with the position accuracy of the ownship symbol. [Chandra, et al. (2003), 6.2.10]
- Text in the pilot's guide and airplane flight manual should document the inaccuracies in the presentation of ownship position and which part of the ownship symbol corresponds to ownship's actual position.
- Loss of ownship positional information should be indicated clearly and immediately.
- If hold lines are provided they should be depicted in their correct relative position.
- If traffic symbols are displayed, text in the pilot's guide and airplane flight manual should document the inaccuracies in the presentation of traffic position and which part of the traffic symbol corresponds to the aircraft's actual position.
- All traffic symbols should be positioned on the display in their appropriate location representative of their actual range.

Suggestion(s)

- Buildings and structures should be displayed in such a manner as to provide information about position, orientation and type. [SAE ARP 5898, 9.2.1.3]

Design Tradeoff(s)

> The depiction of hold lines, runway, taxiway, and ramp area markings will provide the pilot with a greater understanding of ownship position relative to the surface movement area, facilitating visual referencing of out-the-window cues. Unfortunately, the detailed presentation of markings also contributes to clutter and may imply a greater level of accuracy in the system than is actually available. Note that any inaccuracy in the depiction of these markings could make their depiction more misleading than useful.
>
> Similarly, the presentation of buildings on a surface moving map may provide a reference when the pilot is on final approach and provides a means for orientation on the airport surface, but only if the information is accurate and clear. Since airport structures change constantly, keeping an up-to-date accurate database is potentially difficult and costly.

Problem Statement

The depiction of ownship position on maps that are not drawn to scale, or those that are not georeferenced, may be misleading, conveying a sense of greater accuracy than is really present. Some software may only guarantee accuracy of ownship position relative to a surveyed runway position, but more complex displays could accurately depict ownship position relative to other air and ground traffic, taxiways, etc.

The presentation of hold lines may promote a false sense of security. Pilots should be aware of the range of uncertainty regarding the position of aircraft and surface vehicles with respect to hold lines, and should not judge ownship or traffic distance to hold lines based on the information depicted on the surface moving map alone.

The presentation of traffic position on maps that are not drawn to scale, or those that are not georeferenced, may be misleading, conveying a sense of greater accuracy than is really present.

Example(s)

An example error allocation is: position accuracy of 36 meters, data accuracy of 65 meters, latency effects of 4.5 m, display errors of 0.5 m, and aircraft reference point bias of 25 m. The resulting total system accuracy under these assumptions is:

$$[(36 \text{ m})^2 + (65 \text{ m})^2 + (4.5 \text{ m})^2 + (0.5 \text{m})^2]^{1/2} + 25 = 100 \text{ m}$$

These values are only an example, and errors can be allocated differently (e.g., runway database accuracy can be less stringent if the aircraft reference point accuracy error is a smaller value). [TSO C-165/RTCA DO-257A, 2.3.1.1.1]

The circle of uncertainty serves to remind the user that the depicted position is subject to a certain degree of doubt, and further to provide a quantitative indication of this uncertainty.

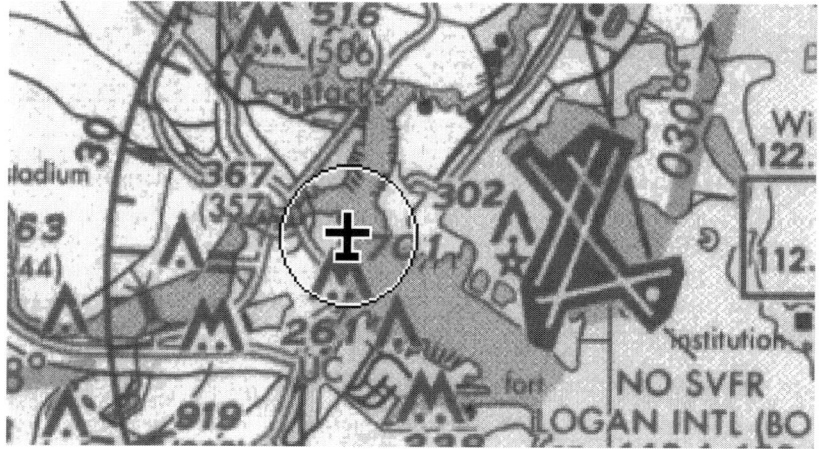

Other alternatives for displaying error include (a) re-sizing ownship depiction so that it is consistent with the total accuracy, or (b) limiting the map range so that the user can not zoom in or out to a range where the data is inaccurate.

The Honeywell and Rockwell Collins displays present a detailed depiction of the airport surface showing not only taxiways and runways, but also the corresponding hold short lines and centerlines.

Evaluation Questions

- Are symbols and graphics positioned so the errors is less than .013 inches on the map depiction or 1% of the shortest axis of the map depiction? Are orientation errors less than 3° with respect to the values provided by the position and database sources?
- Does the display provide an indication when the accuracy implied by the display is better than the level supported by the total system accuracy?
- Is the total system accuracy sufficient for the intended operation?
- Are the inaccuracies in the presentation of ownship position documented in the pilots guide and airplane flight manual? Are inaccuracies in ownship position depicted with a circle of uncertainty around the symbol?
- Is text that indicates the possible inaccuracies in the depiction of airport and other symbols provided in the manual and/or product identification screen?
- Is the aircraft position sensor horizontal positional accuracy for runways less than 36m?
- Is the total database accuracy for runways 43m or less?
- Is the aircraft position sensor horizontal positional accuracy for taxiways less than 36m?
- Is the total database accuracy for taxiways 65m or less?
- Are runway markings depicted in their correct relative position?
- Is ownship position displayed only on georeferenced or to scale maps or charts?
- Is the position accuracy of ownship symbol accurate at all display zoom levels?
- Are pilots provided with information on the limitations of the display of own aircraft position (e.g., pilots manual)? Are pilots provided with information that indicates which part of the ownship symbol corresponds to the ownship's actual position?
- Is loss of ownship position indicated clearly and immediately?
- Are hold lines depicted in their correct relative position?
- If traffic symbols are displayed, are pilots provided with information on the limitations of the display of traffic position? Are pilots provided with information that indicates which part of the traffic symbol corresponds to the aircraft's actual position?
- Are all traffic symbols positioned on the display in their appropriate location and representative of their actual range?

3.3 Ownship

FAA Policy and Guidance

- The surface map display shall contain a symbol representing the location of ownship. [TSO C-165/RTCA DO-257A, 2.3.1.2]
- The ownship symbol shall be unobstructed. [TSO C-165/RTCA DO-257A, 2.2.1.1]

  NOTE: Exceptions may be allowed for multi-function displays depicting higher priority information that are required by regulation that may temporarily obstruct the ownship symbol (e.g., TCAS Traffic Advisory). [TSO C-165/RTCA DO-257A, 2.2.1.1]

- If directional data is available, the ownship symbol should indicate directionality. [TSO C-165/RTCA DO-257A, 2.3.1.2]

  NOTE: A total system accuracy requirement for displaying direction is not defined. The accuracy of the surface moving map is required to be less than 3 degrees of error from the display. The total system accuracy will be evaluated as part of an installation or operational evaluation. For equipment that uses Global Navigation Satellite System source as the position source to derive track, the total system accuracy is expected to be less than 5 degrees when taxiing in a straight line. [TSO C-165/RTCA DO-257A, 2.3.1.2]

- If direction/track is not available, the ownship symbol shall not imply directionality. [TSO C-165/RTCA DO-257A, 2.3.1.2]

  NOTE: When using GNSS track for deriving ownship directionality, if directionality becomes unusable due to low taxi speeds or when stopped, the ownship would revert to a non-directional symbol (e.g., circle). [TSO C-165/RTCA DO-257A, 2.3.1.2]

- If ownship directionality information becomes unusable then this condition should be indicated on the surface map display. [TSO C-165/RTCA DO-257A, 2.3.1.2]
- If the ownship symbol is directional, the front of the symbol that conveys directionality (e.g., apex of a chevron or nose of the aircraft if using an aircraft icon) should correspond to the aircraft location. [TSO C-165/RTCA DO-257A, 2.3.1.2]
- If the ownship symbol is non-directional, the aircraft location should correspond to the center of the non-directional symbol. [TSO C-165/RTCA DO-257A, 2.3.1.2]

Recommendation(s)

- The ownship symbol should be distinct from all other symbology.

Suggestion(s)

- One method for indicating the loss of directionality information may consist of changing the ownship depiction from a directional symbol to a non-directional symbol (e.g., circle). [TSO C-165/RTCA DO-257A, 2.3.1.2]
- Equipment that does not have access to heading information may derive track based on changes in position over time (e.g., a Global Navigation Satellite System (GNSS) sensor used to derive track). However, this information will become unreliable when the taxi speed is low relative to turning velocity. Directionality information is generally considered unusable if it is not within 15 degrees of actual track. [TSO C-165/RTCA DO-257A, 2.3.1.2]
- In order to be consistent with airborne navigation displays, the pilot should have the ability to move ownship from off center to another position (e.g., the center). [SAE ARP 5898, 9.4.3.10]

Design Tradeoff(s)

> One way to make ownship distinct from other symbology is to represent ownship with a unique color. White is a color that is commonly used for ownship as it is stands out against a black background. However, as more display elements are also colored in white – or in colors similar to white, ownship will become less salient. Seems like this belongs as a trade-off.

Problem Statement

The depiction of ownship may promote a false sense of security. While some manufacturers display ownship as larger than true scale in order to reinforce that ownship cannot and should not be used as the basis for maneuvering, special consideration needs to be given to these displays as well as the accuracy of the items depicted.

Example(s)

Examples of ownship representations are presented below.

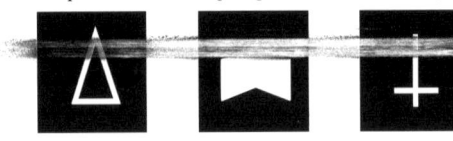

Evaluation Questions

- Does the display contain a symbol representing ownship position?
- Is the depiction of ownship position unobstructed?
- If directional information is available, does ownship symbol show directionality?
- If direction or track information is not available, is ownship represented by a non-directional symbol?
- Is a notification presented when ownship directionality information is unusable?
- If ownship symbol is directional, does aircraft position correspond to the front of the symbol?
- If ownship symbol is non-directional, does aircraft position correspond to the center of the symbol?
- Is the depiction of ownship distinct from all other symbology?

3.4 Runways

FAA Policy and Guidance

> - The capability shall exist to depict runways. [TSO C-165/RTCA DO-257A, 2.3.1.1.1]
> - The depiction of runways shall be distinctive from all other symbology. [TSO C-165/RTCA DO-257A, 2.3.1.1.1]
> - With the exception of instances where two or more runways intersect, each runway should be depicted as a contiguous area (i.e., an unbroken rectangle). [TSO C-165/RTCA DO-257A, 2.3.1.1.1]
> - Runways should be depicted as filled areas, rather than outlined areas. [TSO C-165/RTCA DO-257A, 2.3.1.1.1]

Recommendation(s)

> - When two or more runways intersect, the edges of the intersecting runways should not be drawn through the intersection.
> - Runways should be distinguished from other display elements along a dimension other than color. (See also 2.1 Use of Color)
> - Runways should be depicted with thick solid lines rather than dashed lines so that they will be more salient.

Problem Statement

If runways are not clearly depicted, the pilot will take more time finding the location of runways on the surface map display or may confuse another surface attribute (e.g., a taxiway or ramp area) depicted on the display for a runway. Note that the use of color as the *sole* means of distinguishing runways from other display elements may not be sufficient. The ability to discriminate color quickly and accurately will be problematic for pilots with color deficiencies and inhibited by external factors that influence the perceived color of the display elements (e.g., the illumination on the flight deck or the location of the surface map display). This will result in increased workload and slower response times.

Example(s)

Some manufacturers depict runways schematically by drawing the location of runway centerlines without the corresponding runway edges. This technique reduces the clutter on the display by drawing runways with one line rather than two yet still provides the pilot with a depiction of runway location.

When two runways intersect, the depiction of runway edges through the intersection will give precedence to one runway, as shown in the figure on the left. Depicting the intersection as an unbroken area, as in the figure on the right, will prevent this misperception.

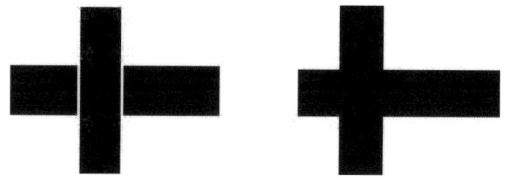

When dashed lines are used to depict runways, runway edges may not be discriminable from the ticks on the compass rose, as the runways approach the edge of the surface display.

One manufacturer ensures the saliency of runways with respect to taxiways by increasing the brightness of runways, so that the contrast between the two is 25% or greater.

Evaluation Questions

- Are runways depicted?
- Is the depiction of runways distinct from other symbology?
- With the exception being when two runways intersect, are runways depicted as a contiguous object?
- Are runways depicted as filled rather than outlined areas?

- When runways intersect, is the depiction of runway edges deleted at the intersection?
- Are runways distinguishable from other display elements by a dimension other than color?
- Are runways depicted with solid lines rather than dashed lines?

3.5 Taxiways
FAA Policy and Guidance

> - The capability should exist to depict taxiways. [TSO C-165/RTCA DO-257A, 2.3.1.1.2]
> - Taxiways should be depicted as filled areas, rather than outlined areas. [TSO C-165/RTCA DO-257A, 2.3.1.1.2]

Recommendation(s)

> - When taxiways intersect runways, the depiction of runways should be given precedence at the intersection.
> - When two or more taxiways intersect, the edges of the intersecting taxiways should not be drawn through the intersection.
> - Taxiways should be clearly depicted through ramp areas.

Problem Statement

If the location of taxiways are not depicted clearly, the pilot may take more time finding the location of taxiways on the surface map display or confuse the depiction of taxiways with other paved areas. The result is increased head-down time, workload, and error.

Example(s)

An information analysis conducted to examine the value of surface attributes for operations on or near the airport surface showed that pilots considered the depiction of taxiways as being of very high value (Yeh and Chandra, 2003a, b).

When taxiways intersect runways, the depiction of runway edges through the intersection will give precedence to the runway. An example is provided in the figure below; the runway is depicted in black and the taxiway in gray.

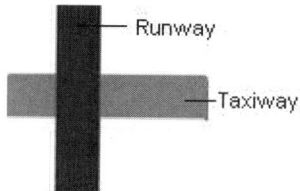

When two taxiways intersect, the depiction of taxiway edges through the intersection will give precedence to one taxiway. Depicting the intersection as an unbroken area, as shown below, will prevent this misperception.

Evaluation Questions
- Are taxiways depicted?
- Are taxiways depicted as filled areas?
- If taxiways intersect runways, is the depiction of runways given precedence at the intersection?
- When taxiways intersect, is the depiction of taxiway edges deleted at the intersection?
- Are taxiway edges through ramp areas clearly indicated?

3.6 Runway/Taxiway Identifiers

FAA Policy and Guidance

- The runway identifiers shall be available for depiction on the display, if available. [TSO C-165/RTCA DO-257A, 2.3.2]
- If taxiways are depicted then the taxiway identifiers should be available for depiction on the display, if available. [TSO C-165/RTCA DO-257A, 2.3.2]
  NOTE: The equipment is not required to continuously display all taxiway and runway identifiers. For example, some implementations may include a de-cluttering function to remove the identifiers. [TSO C-165/RTCA DO-257A, 2.3.2]
- Runway identifiers should be distinguishable from the depiction of runway markings. [TSO C-165/RTCA DO-257A, 2.3.2]
- At reduced map ranges, at least one identifier should be displayed for any taxiway or runway depicted within the selected map range. [TSO C-165/RTCA DO-257A, 2.3.2]
- When surface map features are being depicted, the aerodrome designator (e.g., ICAO identifier) or name for the depicted aerodrome should be indicated on the display. [TSO C-165/RTCA DO-257A, 2.3.2]

Recommendation(s)

- Identifiers should remain upright to facilitate readability.
- Runway identifiers when presented should be legible across all map ranges.

Design Tradeoff(s)

One method of decluttering is to prioritize the presentation of identifiers according to importance relative to ownship's location on the airport surface and taxi route. In this implementation, only the runway and taxiway identifiers relevant to ownship's current path is displayed, hence reducing clutter. However, the effectiveness of this decluttering method still needs to be evaluated to ensure that the removal of identifiers that are not along ownship's taxi route does not diminish the pilots' situation awareness.

Problem Statement

The pilot may not be aware of the runway or taxiway on which s/he is currently taxiing when the display is at lower zoom labels if the labels are spaced at pre-defined intervals, or placed only at runway and/or taxiway intersections. This will increase the potential for errors and lead to increased heads-down time.

Labels that are not properly aligned (e.g., upside down or rotated at some angle) are difficult to read. Text that does not remain upright may lead to confusion, particularly when two numbers or two letters are easy to transpose, e.g., 13/31 or M/W.

Example(s)

The 13/31 runway designation requires extra vigilance by pilots when conducting operations on or near the airport surface, as the numbers are easy to transpose. A runway incursion resulted when a Cessna 172 cleared to land Runway 31L at Boeing Field, landed on Runway 13L after acknowledging Runway 31L. At the same time, a Lancair was landing on Runway 31R. Both aircraft came within 100 feet of each other, before finally coming to a complete stop (June 1, 2002; www.aopa.org/asf/incursions.html).

One method that would allow the presentation of at least one identifier for each runway or taxiway continuously is to vary the distance between labels as a function of map range. At high map ranges, identifiers may be spaced farther apart, whereas at low map ranges, identifiers may be spaced closer together.

The Rockwell Collins PC-based surface display provides a decluttering feature that when enabled, presents only those runway and taxiway labels that are within a limited region surrounding ownship. All other labels are removed.

Evaluation Questions
- Are runway identifiers depicted, if available?
- If taxiways are depicted, are taxiway identifiers depicted?
- Are runway identifiers distinguishable from runway markings?
- Is at least one identifier available for each runway or taxiway that is viewable at low map ranges?
- Is the airport identifier presented when surface map features are depicted?
- Do identifiers remain upright?
- Are runway identifiers legible across all map ranges?

3.7 Hold Lines

Suggestion(s)

> - To reduce clutter, hold lines that are not relevant to the pilot's route may be removed.
> - More advanced implementations should display the hold short lines such that there is a clear distinction between active and inactive lines. Active hold lines are defined as those hold lines that are in effect for the current clearance. Inactive hold lines are defined as those that the ownship is cleared to cross. [SAE ARP 5898, 9.2.2.6]

Problem Statement

The depiction of all the hold lines will result in clutter. There are many hold lines on the airport surface, many of which may be irrelevant to the pilot's assigned route.

Example(s)

The Honeywell and Rockwell Collins displays depict the position of hold short lines with respect to runways and taxiways.

The depicted distance between ownship and the hold short line varies depending on the size of the ownship icon and the zoom level at which the surface moving map is viewed. For example, at high map ranges, ownship position may be depicted so that it appears to have crossed a hold short line when ownship, in fact, has not.

3.8 Non-Movement Areas

Non-movement areas are defined here to be the non-usable areas for aircraft between taxiways, runways, aprons, and/or any combination of the three.

Recommendation(s)

> - The depiction of non-movement areas should be clearly distinguishable from the depiction of movement areas.

Suggestion(s)

> - Non-movement areas should be depicted so that they appear less salient than movement areas.
> - If the system supports operations in and on non-movement areas, all dynamic (movable) objects, obstacles, hazards and environmental features that affect aircraft operation or safety should be depicted if the information is available. [SAE ARP 5898, 9.4.3.4]

Problem Statement

Failure to distinguish non-movement areas from movement areas increases the likelihood that the pilot will erroneously taxi into the non-movement areas.

Example(s)

The results of preliminary research conducted by MITRE suggests that if the surface map is drawn on a black background, the use of darker colors may be preferred to lighter colors in coding grassy and non-movement areas. On a dark background, these darker colors will be less salient than lighter colors (Bone, et al., 2003).

Evaluation Questions

- Is the depiction of non-movement areas distinguishable from the depiction of movement areas along at least one dimension?

3.9 Taxi Route

The term "taxi route" refers to any sequence of taxiway and/or runway fixes (e.g., turn left at Echo) that are interconnected and depict the desired taxi path.

FAA Policy and Guidance

- Taxi route information shall be distinguishable from all other map attributes. [TSO C-165/RTCA DO-257A, 2.3.1.1.3]
- The way taxi routes are depicted in the preview or edit mode shall be distinctive from the depiction of the active taxi route. [TSO C-165/RTCA DO-257A, 2.3.1.1.3]
    NOTE: An active taxi route is defined as the intended path that will be used during taxi. [TSO C-165/RTCA DO-257A, 2.3.1.1.3]
- The depiction of taxi routes should not obscure runway or taxiway identifiers. [TSO C-165/RTCA DO-257A, 2.3.1.1.3]
    NOTE: The intent of this recommendation is to ensure that taxi routes do not completely cover the identifiers. [TSO C-165/RTCA DO-257A, 2.3.1.1.3]

Suggestion(s)

- Taxi routes should be enterable, modifiable and verifiable in a user-friendly manner, by either data link and/or manual entry. They should be formatted so that they can be augmented through automated taxi route assistance (i.e., intelligent route entry system either through datalink or on-board which will minimize workload by reducing the number of required key strokes, particularly when a re-route is given during the taxi phase). When augmented manual inputs are used, the automated function should not override manual inputs without crew alert or a requirement for crew consent. [SAE ARP 5898, 9.2.2.5]
- Taxi routes loaded through a data link should be formatted consistent with onboard systems and should require pilot acknowledgement to execute. [SAE ARP 5898, 9.2.2.5]
- The taxi route could be depicted using a line that has a small width, so as not to obscure runway or taxiway identifiers. If the taxi route is depicted with a line that has a wide width, the fill should be transparent, so that runway and taxiway identifiers are visible.

Design Tradeoff(s)

The presentation of a taxi route could encourage pilots to use the display for control.

The display of taxi route information provides a visual reference of commands from ground control and reduces confusion if changes in the taxi route occurs (Foyle, et al., 1996). However, the presentation of a taxi route introduces several new tasks onto the flight deck. First, taxi route information must be entered into the system. While this is a step that may be automated with the push of a button, in initial implementations of this feature, the crew may be required to enter taxi route information manually. Second, the crew will need to review and check the data for errors, regardless of whether that data has been entered manually or automatically. Finally, the crew may need to edit the route information in real time, e.g., when they receive a re-routing during taxi. It is possible that many of these tasks will need to be performed during pre-take-off and/or final approach preparations, which is already a high workload phase of flight.

Problem Statement

Taxi routes that are depicted poorly may obscure information, e.g., runway or taxiway markings, and/or be mistaken for runways, taxiways, or surface markings resulting in increased head-down time, slower taxi time, and increased potential for errors.

Failure to clearly distinguish taxi routes in preview or edit mode from the active taxi route could result in confusion during taxi.

Example(s)

The NASA Ames surface display (in Figure A-17) shows the taxi route using a line whose thickness is equal to the width of the runway or taxiway. The taxi route is magenta to indicate a cleared route, and white to indicate a pending route. There are several advantages to the NASA Ames implementation. First, the line thickness will prevent pilots from confusing the taxi route with runway or taxiway markings. Second, taxiway labels are drawn on top of the taxi route, so information is not obscured. Finally, the colors used are significantly different from the color used for runways and taxiways which will prevent pilots from confusing the taxi route with runways and taxiways.

Evaluation Questions

- Is taxi route information distinguishable from other map attributes?
- Is the depiction of an active taxi route distinctive from the depiction of the taxi route in preview or edit mode?
- Does the presentation of a taxi route obscure any runway or taxiway labels?

3.10 Prioritization of Map Features

FAA Policy and Guidance

> - To ensure the availability of appropriate information during surface operations, the order of display layer precedence (in case aerodrome features overlap) should be (higher priority layered on top): [TSO C-165/RTCA DO-257A, 2.3.4.1]
>   (a) Ownship symbol
>   (b) Taxi route
>   (c) Runway identifiers
>   (d) Runways
>   (e) Taxiway identifiers
>   (f) Taxiways

Problem Statement

If display elements of high value are obscured, pilots may not know be able to obtain the information they need quickly and accurately, resulting in potential errors and longer head-down time.

Example(s)

Since the airport environment is so rich in detail, it is likely that when all surface attributes are presented, some display elements will be obscured. Each display element on one of Rockwell Collins prototype displays is drawn on an opaque background, so that when airport features overlap, the display element on "top" is discriminable. The display elements are prioritized, so that the more important information is drawn on top. Rockwell Collins defines their prioritization scheme to be:

1. Ownship symbol
2. Other traffic, call sign, flight ID
3. Flight plan
4. Compass rose
5. Range ring
6. Lubber line, trend line
7. Closed runway, NOTAM areas
8. Clearance path
9. Runway designators
10. Runway area
11. Hold areas
12. Closed taxiway NOTAM areas
13. Taxiway designators
14. Taxiway boundaries
15. Ramp designators
16. Service areas
17. Non-movement areas
18. Gate designators
19. Terminal buildings
20. Grassy areas

Evaluation Question(s)

- Is the depiction of surface attributes prioritized so that ownship symbol is unobscured?
- Is the depiction of surface attributes prioritized according to the hierarchy defined in TSO C-165/RTCA DO-257A?

3.11 Indicators (Velocity Vectors, Compass Rose, Lubber Line)

Indicators are display elements that provide reference information on surface moving maps.

Recommendation(s)

- A means to turn the velocity vectors on and off should be provided.
- The units of the horizontal velocity vector should be displayed continuously.
- The units of measurement for the velocity vector should be the same for all displayed traffic and ownship.
- The time value associated with the length of the velocity vector should remain the same when the user zooms in or out of the display.
- Compass rose headings should be labeled, at the least with reference points for north (0°), east (90°), south (180°), and west (270°).

Suggestion

- The lubber line, if available, should be lowlighted, i.e., reduced in intensity, when ownship is on the ground. (see Bone, et al., 2003)
- The lubber line should be available for display.

Design Tradeoff(s)

A compass rose provides an indication of orientation when ownship is in the air and there are few visual references. When ownship is on the ground, however, there is so much detail out-the-window that the depiction of the surface attributes alone may be sufficient for orientation on the surface and the depiction of a compass rose may not be necessary.

The lubber line indicates ownship's future position based on current track. Its depiction may be more useful while ownship is in the air, when track may deviate from current heading, than on the surface, when ownship track and heading are the same. Presentation of a lubber line on the surface may obscure more important information on the display, e.g., taxiway labels or taxi route.

Problem Statement

Units used for data should always be visible in order to prevent confusion and misinterpretation. Formats and units should be consistent for all ownship and traffic, again, to prevent confusion.

Non-standard methods for labeling headings may be confusing if other displays in the flight deck do not use the same convention.

Example(s)

A horizontal velocity vector may be depicted as a straight line extending out in front of a traffic symbol. This vector represents where the traffic will be, based on the current ground speed and direction. An example of an aircraft, indicated here as the chevron, with a velocity vector extending from its nose is presented below.

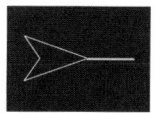

Bone, et al. (2003) reports that pilots commented that the compass rose was not necessary on the surface for taxi operations and could be removed or reduced in intensity. Similar results were reported in a study conducted by Rockwell Collins. Here, pilots turned off the compass rose when viewing a stand-alone CDTI as indications of heading were provided elsewhere (e.g., the navigation display). As a result, in Rockwell Collins' implementation, the compass rose fades into the background when the surface appears.

One convention is to label compass rose headings with "N", "E", "S", and "W".

The Rockwell Collins display shows a lubber line when the map is viewed in heading-up display mode. In track-up or north-up mode, however, the lubber line appears only at the pilot's discretion.

Evaluation Questions

- Is a means to turn the velocity vectors on and off provided?
- Are the units of the horizontal velocity vector displayed continuously?
- Is the unit of measurement for the velocity vector consistent for all displayed traffic and ownship?
- Does the prediction time remain constant when zooming in and out of the display?
- Are compass rose headings labeled?

4 TRAFFIC DISPLAY

This section provides some guidance on the depiction of traffic aircraft and vehicles. While the presentation of traffic information is outside the scope of RTCA DO-257A, the depiction of traffic has been included on surface moving map displays. This guidance is provided in lieu of any published material by aviation authorities.

Note that an Advisory Circular for the *Airworthiness And Operational Approval Considerations For Traffic Surveillance Systems* in currently in development. In the case of any conflicts with that Advisory Circular, the FAA material, as always, takes precedence.

4.1 Traffic Representations

Recommendation(s)

- Each traffic symbol should be positioned at a location representing its relative range and bearing with respect to ownship.
- The traffic symbol should indicate specific directionality, if that data is available and of sufficient quality.
- Surface traffic should be clearly distinguished from airborne traffic. [SAE ARP 5898, 9.4.2.4]
- The crew should be provided a means to select and deselect surface traffic information when appropriate. [SAE ARP 5898, 9.4.2.4] (See also 5.3 Decluttering)

Design Tradeoff(s)

Some systems use different symbols to distinguish traffic from different sensors (e.g., TCAS vs. ADS-B). This provides pilots with an indication of the source of the data but increases the number of symbols the pilot needs to remember.

Providing a depiction of traffic position can reduce the likelihood of a runway incident or accident but the presentation of all traffic on or near the airport surface will result in clutter. (Abbot, et al., 1980)

Problem Statement

Any ambiguity in the representation of traffic position on the surface moving map would increase the heads-down time a pilot needs to obtain the necessary information and the potential for error.

Asking pilots to remember and differentiate between a large number of different symbols will increase workload, increase the likelihood of confusing one symbol for another, and consequently, increase the potential for error.

Example(s)

Inaccuracies in the depiction of traffic position may result in the erroneous depiction of an aircraft on one runway when it is in fact on another or the depiction of an aircraft on a taxiway when it is in fact on a runway.

Presenting unique symbols distinguishing between the different surveillance technologies transmitting traffic positional information increases the number of symbols that may be displayed. In an FAA-sponsored demonstration, pilots noted that at one point, 14+ different icons representing traffic were depicted on the surface map.

Traffic icons used by two manufacturers (Garmin AT and Rockwell Collins) in an FAA-sponsored demonstration are shown below. These symbols sets are provided *not as an endorsement* of a particular symbol set, but rather to highlight the workload issues when a pilot is asked to remember what each symbol represents and differentiate among them. The Garmin AT symbol set has 22 unique symbols; the Rockwell Collins set has 19 unique symbols.

Traffic Display 52

ADS-B-only Traffic	TCAS-only Traffic	ADS-B Traffic Correlated with TCAS Traffic
AVA Target - Airborne	TCAS Other Traffic	AVA Target - Airborne
AVA Target - Airborne	TCAS Proximate Traffic	AVA Target - Airborne
AVA Target - Airborne	TCAS Traffic Alert	ADS-B AVA with TCAS TA
AVA Target - Airborne	TCAS Resolution Advisory	ADS-B AVA with TCAS RA
ADS-B Target - Airborne	TCAS Other Traffic	ADS-B Target - Airborne
ADS-B Target - Airborne	TCAS Proximate Traffic	ADS-B Target - Airborne
ADS-B Target - Airborne	TCAS Traffic Alert	ADS-B Target with TCAS TA
ADS-B Target - Airborne	TCAS Resolution Advisory	ADS-B Target with TCAS RA
Selected ADS-B Target - Airborne	TCAS Other Traffic	Selected ADS-B Target - Airborne
Selected ADS-B Target - Airborne	TCAS Proximate Traffic	Selected ADS-B Target - Airborne
Selected ADS-B Target - Airborne	TCAS Traffic Alert	Selected ADS-B Target with TCAS TA
Selected ADS-B Target - Airborne	TCAS Resolution Advisory	Selected ADS-B Target with TCAS RA
ADS-B Target with CSA alert		
Selected ADS-B Target with CSA Alert		
ADS-B Target - Ground		
Selected ADS-B Target - Ground		
AVA Target - Ground		
Surface Vehicle		
AVA Target - Ground (unknown ground track)		
Fixed Ground or Tethered Obstruction		

ADS-B conflict alerts will not be provided against ADS-B targets correlated with TCAS targets.

No correlation with TCAS targets will be performed on ADS-B ground targets, since the TCAS does not provide information on ground targets.

Garmin AT Symbol set used in a FAA-sponsored demonstration

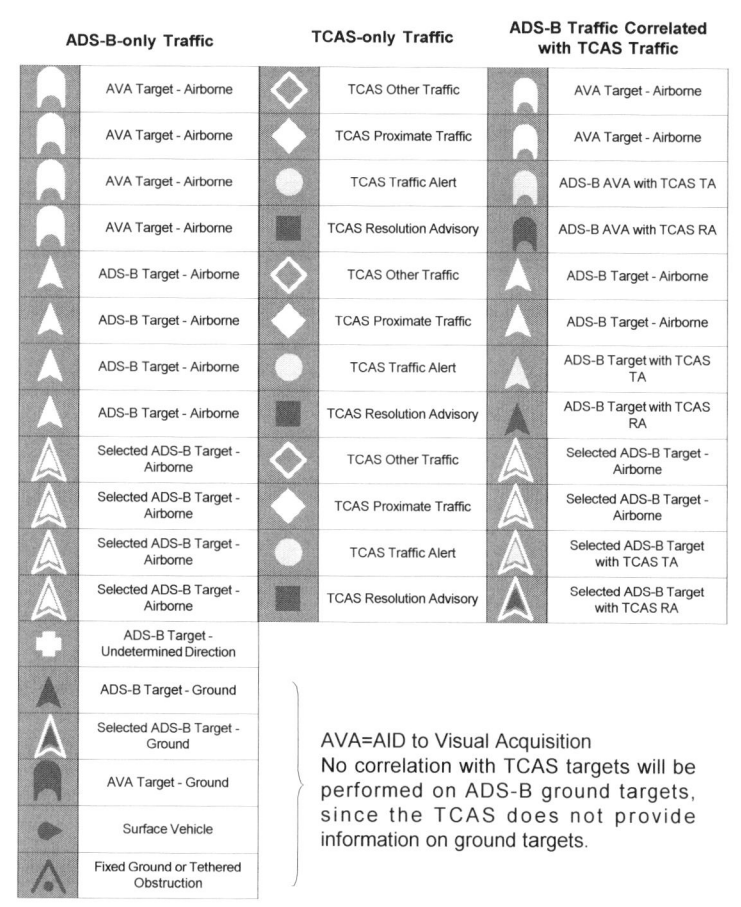

ADS-B-only Traffic		TCAS-only Traffic		ADS-B Traffic Correlated with TCAS Traffic	
⌂	AVA Target - Airborne	◇	TCAS Other Traffic	⌂	AVA Target - Airborne
⌂	AVA Target - Airborne	◆	TCAS Proximate Traffic	⌂	AVA Target - Airborne
⌂	AVA Target - Airborne	○	TCAS Traffic Alert	⌂	ADS-B AVA with TCAS TA
⌂	AVA Target - Airborne	■	TCAS Resolution Advisory	⌂	ADS-B AVA with TCAS RA
▲	ADS-B Target - Airborne	◇	TCAS Other Traffic	▲	ADS-B Target - Airborne
▲	ADS-B Target - Airborne	◆	TCAS Proximate Traffic	▲	ADS-B Target - Airborne
▲	ADS-B Target - Airborne	○	TCAS Traffic Alert	▲	ADS-B Target with TCAS TA
▲	ADS-B Target - Airborne	■	TCAS Resolution Advisory	▲	ADS-B Target with TCAS RA
△	Selected ADS-B Target - Airborne	◇	TCAS Other Traffic	△	Selected ADS-B Target - Airborne
△	Selected ADS-B Target - Airborne	◆	TCAS Proximate Traffic	△	Selected ADS-B Target - Airborne
△	Selected ADS-B Target - Airborne	○	TCAS Traffic Alert	△	Selected ADS-B Target with TCAS TA
△	Selected ADS-B Target - Airborne	■	TCAS Resolution Advisory	△	Selected ADS-B Target with TCAS RA
✦	ADS-B Target - Undetermined Direction				
▲	ADS-B Target - Ground				
△	Selected ADS-B Target - Ground				
▬	AVA Target - Ground				
●	Surface Vehicle				
△	Fixed Ground or Tethered Obstruction				

AVA=AID to Visual Acquisition
No correlation with TCAS targets will be performed on ADS-B ground targets, since the TCAS does not provide information on ground targets.

Rockwell Collins symbol set used in a FAA-sponsored demonstration

Evaluation Questions

- Is each traffic symbol positioned at a location representing its relative range and bearing with respect to ownship position?
- If directionality information is available, does the traffic symbol show directionality?
- Is surface traffic clearly distinguished from airborne traffic?
- Is a method to select/deselect surface traffic provided?

4.2 Selected Traffic

Recommendation(s)

- A control for turning the target selection feature on/off should be provided. [SAE ARP 5898, 9.8.3.3]
- There should be some means of distinguishing the selected aircraft from other traffic.
- Selected aircraft should remain selected until deselected by the user.
- Unselected vehicles should not obscure the selected vehicle unless they are related to a caution or warning.
- The flight crew should be able to select aircraft targets within the currently displayed range. [SAE ARP 5898, 9.8.3.3]

Suggestion(s)

- A representation of the selected aircraft when it is outside the viewable map range may be useful so that the user is aware of its relative bearing.

Design Tradeoff(s)

Selecting or highlighting an aircraft will distinguish it from other traffic aircraft but may also draw the user's attention to the selected aircraft, at the expense of other traffic that are also of importance (Bone, et al. 2003). The salience of the highlighting may need to be evaluated to ensure that it is effective.

One way to select a target aircraft is to click on the aircraft, e.g., with a touch-screen or mouse interface. This provides a direct, intuitive method for target selection, but it may be difficult to select a specific aircraft when there are many aircraft located in close proximity.

Problem Statement

If selected aircraft are not clearly distinguishable from other aircraft, the pilot may not be able to locate the selected aircraft quickly resulting in increased heads-down time and potentially error.

Example(s)

In the Rockwell Collins EFB surface moving map prototype (shown in Figure A-12), turning on the traffic selection function selects the nearest ADS-B/TIS-B aircraft. The selected aircraft is highlighted in cyan. The pilot can then scroll through the other aircraft shown on the display by selecting the **Next** or **Prev** buttons. If the selected aircraft moves outside of the selected map range, an icon appears at the edge of the compass rose to show its relative bearing.

Evaluation Questions

- Can the traffic selection feature be activated and deactivated?
- Is selected aircraft distinguished from other traffic?
- Does the selected aircraft remain selected until deselected by the user?
- Does the selected aircraft remain unobscured by unselected vehicles, unless the unselected vehicle is related to a caution or warning?
- Is the flight crew able to select any aircraft within the currently displayed range?

4.3 Data Blocks/Data Tags

Data tags are located in proximity to the traffic symbol and move with it. *Data blocks* show additional information (e.g., aircraft category) about the target and are placed at a fixed location on the display irrespective of the location of the target.

Recommendation(s)

> - A data tag should have a clear association with the traffic symbol it references.
> - A means should be provided to associate the data block with the traffic symbol.
> - The information presented in data tags should be consistent for all aircraft.
> - An indication should be provided if any piece of information presented in the data field is not available.
> - Traffic identifiers and tags on a display should not obscure each other. The display of tags should be prioritized according to significance to ownship position and route. [SAE ARP 5898, 9.4.3.12]

Suggestion(s)

> - Traffic identifiers/tags should be selectable and deselectable by the flight crew.

Problem Statement(s)

Ambiguous associations between data tags and the aircraft referenced could result in increased heads-down time as the pilot tries to resolve the ambiguity and errors if a pilot associates a data tag with an incorrect aircraft.

Example(s)

In many implementations, when a target vehicle is selected, a data block listing vehicle identification, range, speed, vehicle category, and closure rate is presented in the lower left corner of the display. The text in the data block is presented in the same color as that used to highlight the aircraft. This use of color strengthens the association between the information in the data block and the selected aircraft.

The association between a data tag and the traffic symbol it references can be accomplished through location by presenting the data tag in close proximity to the traffic symbol. However, proximity may not be enough to resolve ambiguity when many aircraft are displayed, e.g., near a busy airport or when viewing the display at high map ranges. In these cases, drawing a leader line to connect the data tag to the traffic symbol may be useful.

One way to indicate that information is not available is to replace its value using a series of dashes, e.g., "- - -", with each dash representing one character.

Evaluation Question(s)

- Is the association between a data tag and the traffic symbol it references clear?
- Is the association between a data block and the traffic symbol it references clear?
- Is the information presented in all data tags consistent for all aircraft?
- Is an indication provided if any information presented in the data field is not available?
- Are traffic identifiers and tags displayed so that they do not obscure one another? Are data tags prioritized according to significance to ownship position and route?

4.4 Altitude Representations

Recommendation(s)

> - A capability to select an altitude band within which to display traffic should be provided to the flight crew.
> - If an aircraft creates an alert situation, then that aircraft should be displayed, regardless of the setting of the altitude filter.
> - If the crew can filter the display of traffic based on the altitude of the aircraft, the altitude filter setting should be continuously displayed.
> - The altitude filter should have a pre-set minimum to ensure that the pilot does not accidentally filter out all traffic.
> - Altitude values should be displayed for airborne traffic. If the traffic altitude is not available, an indication that it is not available should be displayed (e.g., NO ALT in the data tag).
> - Altitudes for traffic simultaneously displayed should be consistent, all altitudes being displayed either in absolute or relative terms.
> - The display should indicate whether absolute or relative altitude is displayed.
> - The display should indicate whether traffic is above or below own-ship.
> - Altitude values should be inhibited for vehicles on the ground.

Problem Statement

If the altitude filter setting is not indicated, the pilot could be surprised if an aircraft that is not displayed creates an alert situation and fail to respond appropriately.

Inconsistencies in the presentation of traffic altitude will be confusing and could lead to errors.

Example(s)

The TCAS convention for indicating whether traffic is above or below own-ship is to show the traffic altitude value above or below the traffic symbol. One method is to use a "+" symbol to designate traffic above ownship position and the "-" symbol to designate traffic below ownship position; the altitude for traffic at the same altitude as ownship may be displayed as "00". The two methods may be combined so that the text is presented above or below the traffic symbol to indicate redundantly whether the traffic is above or below ownship respectively.

The representation of an aircraft 3600 feet above ownship position may be labeled as +36.

Evaluation Questions

- Is a capability to select an altitude band within which to display traffic provided?
- Are aircraft that create alert situations displayed, regardless of the altitude filter setting?
- If the crew can selectively filter altitude, is the selected altitude band continuously displayed?
- Does the altitude filter have a pre-set minimum to prevent the pilot from filtering out all traffic?
- Are altitude values displayed for all traffic in the air? If altitude is not available for a specific aircraft, is an indication that it is not available provided?
- Is the presentation of altitude (i.e., whether it is relative or absolute) consistent for all aircraft?
- Does the display indicate whether actual or relative altitude is displayed?
- Does the display indicate whether the traffic aircraft is above or below ownship?
- Are altitude values inhibited for surface vehicles?

5 FUNCTIONALITY

This section provides considerations to improve the usability of functions available on surface moving map displays.

5.1 Map Range and Panning

FAA Policy and Guidance

- The display shall have the capability of manually changing the map range. [TSO C-165/RTCA DO-257A, 2.2.4]
- Current map range shall be indicated continuously. [TSO C-165/RTCA DO-257A, 2.2.4]
- The electronic map display should provide an indication if the map range is smaller (i.e., "zoomed in" closer) than the level supported by the accuracy and resolution of the data. [TSO C-165/RTCA DO-257A, 2.2.1]
- When using the panning and/or range selection function, an indicator of ownship current position within the overall displayed image should be provided. [TSO C-165/RTCA DO-257A, 2.2.4]
- If a panning and/or range selection function is available, the equipment should provide the capability to return to an ownship-oriented display with a maximum of two discrete control actions (e.g., two button pushes). [TSO C-165/RTCA DO-257A, 2.2.4]

  NOTE: The panning function is defined as moving the center reference of the display independent of ownship. [TSO C-165/RTCA DO-257A, 2.2.4]

Recommendation(s)

- If map range is changed via discrete controls (e.g., buttons or key presses), then separate controls should be provided for increasing and decreasing map range.
- Range markings within the range arc should be labeled.

Suggestion(s)

- If range up and range down buttons are used they should be located next to each other. [SAE ARP 5898, 9.8.3.1]
- If there is a zoom feature, there should also be a coordinated de-clutter feature so that the display remains usable when it is zoomed out (i.e., shows a large area). If there is no de-cluttering when a large area is in view, there may be so many small objects on the display that none of the information is useful. [Chandra, et al. (2003), 6.2.11]
- The minimum ranges for the surface moving map may be set at the points where any decrease in map range would render the information unusable.
- A range selection feature, if installed, on the surface display may be accomplished with a rotary dial or lever rather than buttons that cycle up and down the ranges. This implementation allows the pilot to select the lowest useable range without looking at the display controls by simply rotating the knob to the left until the stop is reached. Cycling through ranges with range up and range down buttons requires more heads down time. [SAE ARP 5898, 9.8.3.1]

Problem Statement

If map range is not clearly indicated, pilots will not be able to determine map range easily, leading to increased workload, head-down time, and errors. Determining the range of the map display will be difficult if hash marks provided for reference are not labeled or have labels that are not clearly associated with range.

Zooming in to low map ranges (e.g., ¼ mile range) may implicitly convey a greater level of data reliability and integrity than what is supported.

A consequence of being zoomed in at a close level is that the pilot could lose awareness of proximal aircraft, which may be out of the current viewing area. An alerting scheme could be implemented to notify pilots of this threat.

While panning across a map, moving from one area to another, the operator may lose track of what is being displayed, and be uncertain how to move in order to see some other area of interest. An indicator of current position may help operators to maintain overall orientation (Foley and van Dam 1982).

Example(s)

The implementation of zooming may be combined with decluttering so that the display elements used to describe the airport surface vary depending on the zoom level. As map range decreases, the airport surface could be described in more detail, and vice versa. Display elements that might not be legible at higher zoom ranges (e.g., non movement areas, taxiway labels) would not be presented until the surface moving map was configured to a range where such data presentation would be supported.

The results of an informal study with two-pilot crews, using a primary flight display and navigation display similar to Boeing 777 or 747-400 displays, indicated that the most commonly used zoom ranges were ½ mile and 1 mile. The pilot flying typically used ½ mile range whereas the pilot not flying selected a more global view of the surface using a 1 mile range (Bone, et al., 2003).

As shown below, each range marking (indicated by the "+") is labeled with the range depicted at that point, as recommended.

Evaluation Questions

- Can the map range be changed manually?
- Is current map range indicated continuously?
- Is an indication provided if the map range is smaller than the level supported by the accuracy and resolution of the data?
- Is an indication of ownship's current position within the overall displayed image provided while panning and/or zooming?
- If panning and/or zooming is available, can the user return the map to an ownship-oriented display after panning or zooming in two discrete control actions or less?
- If map range is changed via discrete controls, are separate controls provided (i.e., one for zooming in and the other for zooming out)?
- Are all range markings within the range arc labeled?

5.2 Autozoom

FAA Policy and Guidance

> - If the display is controlling the map range automatically, the mode (e.g., auto map range) should be indicated. [TSO C-165/RTCA DO-257A, 2.2.4]
> - If the automatic map range function is deactivated, the display should maintain the last range scale prior to deactivation until the flight crew manually selects another map range. [TSO C-165/RTCA DO-257A, 2.2.4]
> - If the display is controlling the map range automatically, then the capability shall exist to activate or deactivate the automatic map range. [TSO C-165/RTCA DO-257A, 2.2.4]
>
>   NOTE: An acceptable method of compliance is to have a discrete control action (e.g., button push) to activate the automatic range function. [TSO C-165/RTCA DO-257A, 2.2.4]

Recommendation(s)

> - Changes in map range should be obvious to the user and should not contribute to mode confusion.

Suggestion(s)

> - The automatic range function may be activated by a discrete control action (e.g., button push).

Design Tradeoff(s)

> In defining an autozoom algorithm, it is important to consider which display elements to present and when. The minimum set of display elements to be displayed has not been determined.

Problem Statement

A clear indication of map range is especially important when the autozoom function is available because changes in map range will occur without the pilot's knowledge. Pilots distribute attention between the surface moving map, other flight displays, and the out-the-window view, without focusing on any one display for a significant length of time.

Example(s)

The autozoom function configures the view of the runway environment during final approach so that the flight crew does not need to interact with the surface moving map during this high workload phase of flight. Proposed methods for autozoom include associating changes in map range to aircraft speed, mechanics (i.e., weight on wheels), altitude, range to runway, phase of flight, or pilot action.

One way that autozoom has been designed, as defined by Garmin AT and MITRE, is to adjust map range on final approach so that runways are continuously in sight. When autozoom is enabled, runways automatically appear when ownship is 8 miles from the airport. As ownship approaches the runway, the distance from ownship position to the four edges of the designated runway is calculated, and the map is zoomed in so that the end of the designated runway remains at the top of the display. Taxiways appear when ownship is 750 feet away. The rest of the surface attributes are presented when ownship touches down.

Evaluation Questions

- If map range is controlled automatically, is this mode indicated?
- If the automatic map range function is deactivated, does the display maintain the last map range, prior to deactivation?
- Is functionality provided to activate and deactivate the automatic map range function?
- Are changes in map range obvious to the user?

5.3 Decluttering

Note: See also 2.14 Shared Display Considerations and 4.1 Traffic Representations.

FAA Policy and Guidance

> The surface moving map may display, either continuously or selectively, information beyond the minimum required data set, defined in RTCA DO-257A[*]. If additional information is displayed on the surface moving map (beyond that required in RTCA DO-257A), then the following requirements apply.
> - The display shall have the capability for manual de-cluttering during operational use. [TSO C-165/RTCA DO-257A, 2.2.1.3]
> - If additional map information has been selected for display, it should be possible to deselect all displayed additional information as a set. [TSO C-165/RTCA DO-257A, 2.2.1.3]
> - It should be possible for the pilot to accomplish this de-clutter function with a single action. [TSO C-165/RTCA DO-257A, 2.2.1.3]

Recommendation(s)

- The decluttering scheme should be documented in the pilots' guide and in the certification plan.
- If there is a de-clutter capability, it should not be possible for the pilot to remove safety critical display elements (e.g., terrain, obstructions, or special use airspace) without knowing that they are suppressed. If such information can be de-cluttered, it should not be possible for the pilot to believe that it is not visible because it is not there. [Chandra, et al. (2003), 6.2.11]
- Managing the display configuration (e.g., scale, orientation, and other options and settings) should not induce significant levels of workload. That is, routine display configuration changes should be minimized. [Chandra, et al. (2003), 6.2.11]
- Managing the display configuration should not result in a significant increase in head down time nor take attention away from other tasks for extended periods of time.
- The implementation of any decluttering scheme must be validated to ensure that the display elements are organized in such a way where the information available is usable to the pilot.

Suggestion(s)

- Decluttering algorithms should de-clutter in the order of least to most significant information.
- Proximate and hazardous vehicles and obstacles should be the last to be de-cluttered. [SAE ARP 5898, 9.4.3.13]

Design Tradeoff(s)

Manual control over the display of individual information elements (e.g., obstructions, or navigational aids) could become complex. On the other hand, automatic display configuration could be frustrating to use, and potentially unsafe, if it does not match the user's expectations. [Chandra, et al. (2003), 6.2.11]

The ability to declutter and/or customize the display provides an alternative to complete removal of levels of information; display decluttering facilitates information search and readability relative to displays that contain clutter (Hofer, Palen, and Possolo, 1993; Mykityshyn, Kuchar, and Hansman, 1994; Schultz, 1986). However, the flexibility to declutter the display (or deciding whether or not decluttering is required) imposes a time cost in retrieving the information (Yeh and Wickens, 2001).

---

[*] The minimum required data set for surface moving map displays are ownship and runways.

Problem Statement
> Too much flexibility, e.g., having many options and levels of decluttering, could increase the time it takes the pilot to select the appropriate level of decluttering.

Example(s)
> One method for decluttering is to present taxiway position schematically, e.g., by depicting taxiway centerlines rather than the taxiway edges. This reduces clutter on the surface moving map display, but still provides the pilot with a general taxiway layout.

Evaluation Questions

- Is manual decluttering available?
- If additional information beyond the minimum required data set defined in RTCA DO-257A is displayed, is it possible to deselect all additional information as a set?
- Can decluttering be accomplished in a single action?
- Is the decluttering scheme documented in the pilots' guide and certification plan?
- Is decluttering implemented in such a way that alerts and other safety information can *not* be decluttered?
- Is decluttering implemented in a way that it does not induce significant levels of workload?
- Is decluttering implemented in a way that managing the display configuration does not significantly increase head down time nor take attention away from other tasks for an extended period of time?
- Was the decluttering scheme validated?

5.4 Map Orientation

FAA Policy and Guidance

- Current map orientation shall be clearly, continuously, and unambiguously indicated (e.g., track-up vs. North-up). [TSO C-165/RTCA DO-257A, 2.2.4]

  NOTES: [TSO C-165/RTCA DO-257A, 2.2.4]

  1. Issue: systems exist that have four orientation modes available without any explicit indication of mode: actual track-up, North-up, heading-up, desired track-up. The orientation mode selected must be continuously indicated. Alternatively, the indication could be done using external annunciators or an external switch that indicates the orientation currently selected.

  2. An acceptable means of compliance would be to have a "desired track-up" (or DTK↑), "North-up" (or N ↑), "heading-up" (or HDG ↑) or "actual track-up" (or TRK ↑) on the display.

  3. A compass arc/rose or North indicator is an acceptable means of compliance for a system that provides only two options (North-up and one other option).

- The display shall have the capability to present map information in at least one of the following orientations: actual track-up or heading-up. [TSO C-165/RTCA DO-257A, 2.2.4]

  NOTES: [TSO C-165/RTCA DO-257A, 2.2.4]

  1. In addition to the above, desired track-up and North-up orientations (to facilitate cross checking with the paper charts and flight planning) are also acceptable.

  2. Default of track-up or heading-up in-flight is encouraged.

  3. This requirement does not apply to systems while displaying RAC data.

- If desired track-up orientation is used, the aircraft symbol shall be oriented to actual track or heading. [TSO C-165/RTCA DO-257A, 2.2.4]

- If the flight crew has selected a display orientation (e.g., track-up), that display orientation should be maintained until an action that requires an orientation change occurs. [TSO C-165/RTCA DO-257A, 2.2.4]

  NOTE: Actions can include crew selection of a different orientation or a mode change (e.g., TCAS auto popup). [TSO C-165/RTCA DO-257A, 2.2.4]

- If the system is in North-up, the orientation of the map shall be referenced to true North. [TSO C-165/RTCA DO-257A, 2.2.4]

- In desired track-up orientation, it is recommended that a track extension line that projects the actual track out from the aircraft be displayed. [TSO C-165/RTCA DO-257A, 2.2.4]

- Consideration should be given to the potential for confusion that could result from presentation of relative directions (e.g., positions of other aircraft on traffic displays) when the display is positioned in an orientation inconsistent with that information. For example, it may be misleading if own aircraft heading is pointed to the top of the display and the display is not aligned with the aircraft longitudinal axis. [AC 120-76A, 10.b(3)]

Design Tradeoff(s)

The surface moving map may have a north-up, track-up, or heading-up orientation. Each orientation is suitable for a different situation. A north-up orientation is beneficial for planning when the pilot is cross checking the paper charts. A heading-up or track-up display is beneficial for tactical decisions, e.g., determining which way to turn from the runway, as it reduces the need for mental rotation since the information shown on the map corresponds to the view out-the-window.

Problem Statement

If the current orientation of the map is not clearly indicated, the pilot may confuse the orientation modes and end up following an incorrect path and/or turn towards a landmark that is not really there.

Example(s)

> An acceptable means of indicating map orientation would be to have a "desired track-up" (or DTK↑), "North-up" (or N ↑), "heading-up" (or HDG ↑) or "actual track-up" (or TRK ↑) on the display. [TSO C-165/RTCA DO-257A]
>
> A compass arc/rose or North indicator is an acceptable means of indicating map orientation for a system that provides only two options (North-up and one other option). [TSO C-165/RTCA DO-257A]

Evaluation Questions

- Is the current map orientation clearly, continuously, and unambiguously indicated?
- Can map information be presented in either a track-up or heading-up orientation?
- If track-up orientation is used, is ownship symbol oriented to the aircraft's track or heading?
- Is the map orientation maintained until it is changed by the crew?
- If North-up orientation is selected, is the orientation of the map referenced to true North?
- If a track-up orientation is presented, is a track extension line that projects the actual track out from the aircraft displayed?
- Is the surface map display equipment installed so that the track- or heading-up display is aligned properly within the flight deck?

REFERENCES

FEDERAL AVIATION ADMINISTRATION (FAA) PUBLICATIONS:

Title 14 of the Code of Federal Regulations (CFR) 23, *Airworthiness Standards: Normal, Utility, Acrobatic, and Commuter Category Airplanes.*

Title 14 of the Code of Federal Regulations (CFR) 23.771, *Pilot Compartment.*

Title 14 of the Code of Federal Regulations (CFR) 23.773, *Pilot Compartment View.*

Title 14 of the Code of Federal Regulations (CFR) 23.777, *Cockpit Controls.*

Title 14 of the Code of Federal Regulations (CFR) 23.779, *Motion and Effect of Cockpit Controls.*

Title 14 of the Code of Federal Regulations (CFR) 23.1301, *Function and Installation.*

Title 14 of the Code of Federal Regulations (CFR) 23.1309, *Equipment, Systems, and Installations.*

Title 14 of the Code of Federal Regulations (CFR) 23.1321, *Arrangement and Visibility.*

Title 14 of the Code of Federal Regulations (CFR) 23.1322, *Warning, Caution, And Advisory Lights.*

Title 14 of the Code of Federal Regulations (CFR) 23.1523, *Minimum Flight Crew.*

Title 14 of the Code of Federal Regulations (CFR) 23.1543, *Instrument Markings: General.*

Title 14 of the Code of Federal Regulations (CFR) 23.1555, *Control Markings.*

Title 14 of the Code of Federal Regulations (CFR) 25.771, *Pilot Compartment.*

Title 14 of the Code of Federal Regulations (CFR) 25.773, *Pilot Compartment View.*

Title 14 of the Code of Federal Regulations (CFR) 25.777, *Cockpit Controls.*

Title 14 of the Code of Federal Regulations (CFR) 25.1301, *Function and Installation.*

Title 14 of the Code of Federal Regulations (CFR) 25.1309, *Equipment, Systems, and Installations.*

Title 14 of the Code of Federal Regulations (CFR) 25.1321, *Arrangement and Visibility.*

Title 14 of the Code of Federal Regulations (CFR) 25.1322, *Warning, Caution, And Advisory Lights.*

Title 14 of the Code of Federal Regulations (CFR) 25.1523. *Minimum Flight Crew.*

Title 14 of the Code of Federal Regulations (CFR) 25.1543, *Instrument Markings: General.*

Title 14 of the Code of Federal Regulations (CFR) 25.1555, *Control Markings*

Title 14 of the Code of Federal Regulations (CFR) 27.771, *Pilot Compartment.*

Title 14 of the Code of Federal Regulations (CFR) 27.773, *Pilot Compartment View.*

Title 14 of the Code of Federal Regulations (CFR) 27.777, *Cockpit Controls.*

Title 14 of the Code of Federal Regulations (CFR) 27.1301, *Function and Installation.*

Title 14 of the Code of Federal Regulations (CFR) 27.1309, *Equipment, Systems, and Installations.*

Title 14 of the Code of Federal Regulations (CFR) 27.1321, *Arrangement and Visibility.*

Title 14 of the Code of Federal Regulations (CFR) 27.1322, *Warning, Caution, And Advisory Lights.*

Title 14 of the Code of Federal Regulations (CFR) 27.1523. *Minimum Flight Crew.*

Title 14 of the Code of Federal Regulations (CFR) 27.1543, *Instrument Markings: General.*

Title 14 of the Code of Federal Regulations (CFR) 27.1555, *Control Markings*

Title 14 of the Code of Federal Regulations (CFR) 29.771, *Pilot Compartment.*

Title 14 of the Code of Federal Regulations (CFR) 29.773, *Pilot Compartment View.*

Title 14 of the Code of Federal Regulations (CFR) 29.777, *Cockpit Controls.*

Title 14 of the Code of Federal Regulations (CFR) 29.1301, *Function and Installation.*

Title 14 of the Code of Federal Regulations (CFR) 29.1321, *Arrangement and Visibility.*

Title 14 of the Code of Federal Regulations (CFR) 29.1322, *Warning, Caution, And Advisory Lights.*

Title 14 of the Code of Federal Regulations (CFR) 29.1523. *Minimum Flight Crew.*

Title 14 of the Code of Federal Regulations (CFR) 29.1543, *Instrument Markings: General.*

FAA Advisory Circular (AC) 20-131A, *Airworthiness Approval of Traffic Alert and Collision Avoidance Systems (TCAS II) and Mode S Transponders.*

FAA Advisory Circular (AC) 23.1311-1A, *Installation of Electronic Displays In Part 23 Airplanes.*

FAA Advisory Circular (AC) 25-11, *Transport Category Airplane Electronic Display Systems.*

FAA Advisory Circular (AC) 120-74, *Part 121, 125, and 135 Flightcrew Procedures During Taxi Operations.*

FAA Safe Flight 21. *Operational Evaluation-2. Final Report.*

FAA Technical Standard Order (TSO)-C63c, Airborne Weather And Ground Mapping Pulsed Radars.

FAA Technical Standard Order (TSO)-C105, Optional Display Equipment For Weather And Ground Mapping Radar Indicators.

FAA Technical Standard Order (TSO)-C110a, Airborne Passive Thunderstorm Detection Equipment.

FAA Technical Standard Order (TSO)-C113, Airborne Multipurpose Electronic Displays.

FAA Technical Standard Order (TSO)-C118, Traffic Alert And Collision Avoidance System (TCAS) Airborne Equipment, TCAS I.

FAA Technical Standard Order (TSO)-C119b, Traffic Alert and Collision Avoidance Systems (TCAS) Airborne Equipment, TCAS II.

FAA Technical Standard Order (TSO)-C129a, Airborne Supplemental Navigation Equipment Using the Global Positioning System (GPS).

FAA Technical Standard Order (TSO)-C147, Traffic Advisory System (TAS) Airborne Equipment.

FAA Technical Standard Order (TSO)-C151b, Terrain Awareness and Warning System (TAWS).

FAA Technical Standard Order (TSO)-C165, Electronic Map Display Equipment for Graphical Depiction of Aircraft Position.

DEPARTMENT OF TRANSPORTATION (DOT) PUBLICATIONS:
DOT/FAA/CT-03/05, *Human Factors Design Standard for Acquisition of Commercial-off-the-shelf Subsystems, Non-Developmental Items, and Developmental Systems* (V. Ahlstrom and K. Longo, 2003). Atlantic City, NJ: Federal Aviation Administration Technical Center. Available at http://hf.tc.faa.gov/hfds/

INTERNATIONAL CIVIL AVIATION ORGANIZATION (ICAO) PUBLICATIONS:
ICAO Annex 4. *Aeronautical Charts*, Annex 4 to the Convention on International Civil Aviation, 10th edition, July 2001.

ICAO Annex 15. World Geodetic System-1984 (WGS-84), *International Standards and Recommended Practices, Aeronautical Information Services*, Annex 15 to the Convention on International Civil Aviation, 10th Edition, July 1997.

ICAO 8400/5. *Procedures for Air Navigation Services ICAO Abbreviations and Codes.* Fifth Edition-1999.

INTERNATIONAL ORGANIZATION FOR STANDARDIZATION (ISO) PUBLICATIONS:
ISO 9186. *Graphical symbols – Test methods for judged comprehensibility and for comprehension.* Second edition, 2001.

JOINT AVIATION AUTHORITIES (JAA) PUBLICATIONS:
Advisory Material Joint (AMJ) 25-11. *Electronic Display Systems.*

Advisory Material Joint (AMJ) 25.1322. *Alerting Systems.*

MILITARY PUBLICATIONS:
DOD, *GPS Standard Positioning Service Performance Standard*, October 2001.

Military Standard 1472D (1989), *Human Engineering Design Criteria For Military Systems, Equipment, And Facilities.* Washington, DC: U.S. Department of Defense.

NAWCADPAX-96-268-TM *Situational Awareness Guidelines.*

RTCA, INC. DOCUMENTS
RTCA DO-200A/EUROCAE ED-76, *Standards for Processing Aeronautical Data.*

RTCA DO-236A, *Minimum Aviation System Performance Standards: Required Navigation Performance for Area Navigation.*

RTCA DO-257A, *Minimum Operational Performance Standards for the Depiction of Navigational Information on Electronic Maps.*

RTCA DO-267, Minimum Aviation System Standards (MASPS) for Flight Information Services Broadcast (FIS-B) Data Link.

RTCA DO-272/EUROCAE ED-99, User Requirements for Aerodrome Mapping Information.

SOCIETY OF AUTOMOTIVE ENGINEER (SAE) PUBLICATIONS
SAE AIR 1093, *Numeral, Letter and Symbol Dimensions for Aircraft Instrument Displays.*

SAE ARP 4032, *Human Engineering Considerations in the Application of Color to Electronic Aircraft Displays.*

SAE ARP 4101/2, *Pilot Visibility from the Flight Deck.*

SAE ARP 4102, *Flight Deck Panels, Controls, and Displays.*

SAE ARP 4102/4, *Flight Deck Alerting Systems.*

SAE ARP 4102/7, *Electronic Displays*: and Appendix A, B, & C.

SAE ARP 5898, *Human Interface Criteria for Flight Deck Surface Operations Displays.*

OTHER PUBLICATIONS:
Battiste, V., Downs, M., Sullivan, B.T., and Soukup, P.A. (1996a). Utilization of a ground taxi map to support low visibility surface operations. *Proceedings of the AIAA Flight Simulation Technologies Conference*, 393-403.

Battiste, V., Downs, M., and McCann, R.S. (1996b). Advanced taxi map display design for low-visibility operations. *Proceedings of the Human Factors and Ergonomics Society 40th Annual Meeting* (pp. 997-1001)

Chandra, D.C., Yeh, M., Riley, V., and Mangold, S.J. (2003). *Human Factors Considerations in the Design and Evaluation of Electronic Flight Bags (EFBs). Version 2.* DOT-VNTSC-FAA-03-07 and DOT/FAA/AR-03/67. Cambridge, MA: USDOT Volpe Center. Available at www.volpe.dot.gov/opsad/efb.

Coyle, S. (June 5, 1997). *Aircraft On-Board Navigation Data Integrity – A Serious Problem*, Transport Canada Database Working Group Paper, Office of Safety and Security: Ottowa, Canada.

Foley, J.D. and van Dam, A. (1982). *Fundamentals of Interactive Computer Graphics*. Addison-Wesley: Reading, MA.

Foyle, D.C., Andre, A.D., McCann, R.S., Wenzel, E.M., Begault, D.R., and Battiste, V. (1996). Taxiway navigation and situation awareness (T-NASA) system: Problem, design philosophy, and description of an integrated display suite for low-visibility airport surface operations. *Proceedings of the SAE/AIAA World Aviation Congress*.

Hofer, E., Palen, L, and Possolo, A. (1993). Flight Deck Information Management: Functional Integration of Electronic Approach Information. *Seventh International Symposium on Aviation Psychology*, Columbus OH.

Hooey, B. L., Foyle, D. C., & Andre, A. D. (2000). Integration of cockpit displays for surface operations: The final stage of a human-centered design approach . *SAE Transactions: Journal of Aerospace*, *109*, 1053-1065.

Mykityshyn, M.G., Kuchar, J.K, and Hansman, R.J. (1994). Experimental study of electronically based instrument approach plates. *International Journal of Aviation Psychology*. 4(2), 141-166.

Sanders, M.S. and McCormick, E.J. (1993). *Human Factors in Engineering and Design*. McGraw-Hill: New York.

Sharp, E. and Hornseth, J.P. (1965). *The effects of control location upon performance time for know, toggle switch, and push button* (Technical Report. 65-41). AMRL. Wright-Patterson Air Force Base, OH.

Siegel, A.I., and Brown, F.R. (1958). An experimental study of control console design. *Ergonomics*, 1, 251-257.

Wreggit, S. S., & Marsh, D. K. II. (1998). Cockpit integration of GPS: Initial assessment—menu formats and procedures (DOT/FAA/AM-98/9). Washington, DC: Federal Administration Office of Aviation Medicine.

Yeh, M. and Chandra, D. (2003a). Air Transport Information Priorities for Surface Moving Maps. *Proceedings of the 47th Annual Meeting of the Human Factors and Ergonomics Society*. Human Factors and Ergonomics Society: Santa Monica, CA.

Yeh, M. and Chandra, D. (2003b). Determining Minimal Display Element Requirements for Surface Map Displays. *Proceedings of the 12th International Symposium on Aviation Psychology*. Wright State University: Dayton, OH.

Yeh, M. and Wickens, C.D. (2001). Attentional filtering in the design of electronic map displays: A comparison of color-coding, intensity coding, and decluttering techniques. *Human Factors*, 43(4), 543-562.

APPENDIX A: INDUSTRY OVERVIEW

An industry overview of moving map displays that depict ownship position on the airport surface was conducted to provide a snapshot as to the state of the display development. The opportunity to view these prototype and FAA-approved displays in development by vendors and research organizations occurred through invitations to vendor sites or through demonstrations or descriptions at public forums (e.g., technical meetings or FAA-sponsored demonstrations). Note that as with any software development cycle, changes in the display design occur frequently; as a result, accuracy of the information cannot be guaranteed.

The displays listed in Tables A-1 and A-2 were available for review. Table A-1 lists vendor displays, and Table A-2 describes displays being used by research organizations. Additional information on these displays may be found in the corresponding sections. For each display, a brief description of the prototype is provided with a table listing the display elements depicted, the method of depiction, and the proposed methods for interaction. The level of detail with which each prototype is described varies as a function of the level of opportunity to view and interact with the display. Images are provided when available.

The following manufacturers have made the information on their surface moving map displays publicly available. It is important to note that these are not the only manufacturers developing surface moving map displays.

Vendor	Sec	Product	Status	Website
Aviation Communication & Surveillance Systems (ACSS) Phoenix, AZ	A.1.1		Prototype	www.l-3com.com/acss/
Diehl Avionics Frankfurt, Germany	A.1.2		Prototype	www.diehl.com
Garmin Advanced Technologies (AT) Salem, OR	A.1.3.1	MX-20*	TSO	www.garminat.com/mx20_gen.shtml
	A.1.3.2	AT2000*	TSO	www.garminat.com/at2000_air.shtml
Honeywell Phoenix, AZ	A.1.4		Prototype	www.honeywellaerospace.com
Jeppesen Denver, CO	A.1.5.1	Jepp View Flite Deck		www.jeppesen.com
	A.1.5.2	Taxi Position Awareness	Prototype	
Rockwell Collins, Inc. Cedar Rapids, IA	A.1.6.1	Cockpit display	Prototype	www.rockwellcollins.com
	A.1.6.2	EFB display	Prototype	
Smiths Aerospace Grand Rapids, MI	A.1.7	Flight Management Computer System (FMCS)	Prototype	www.smithsind-aerospace.com

Table A-1. Industry vendors (* = FAA-approved display). Information updated August, 2003.

At this time, the Garmin AT MX20, AT2000, and Jeppesen Taxi Position Awareness displays have obtained the appropriate TSO's. Note that the FAA has developed TSO-C165, "Electronic Map Display Equipment for Graphical Depiction of Aircraft Position."

The following research organizations have been interested in empirically examining the benefits of surface moving map displays and have developed their own prototype displays for these purposes.

Research Organization	Sec	Website
William J. Hughes Technical Center (FAA) Atlantic City, NJ	A.2.1	www.tc.faa.gov
MITRE McLean, VA	A.2.2	www.caasd.org
NASA-Ames Moffet Field, CA	A.2.3	human-factors.arc.nasa.gov/ihi/hcsl/T-NASA.html

Table A-2. Research organizations.

More information about these prototype displays can be found in the corresponding sections listed below. Note that since these prototypes are developed for research purposes only, none of these displays are FAA approved.

The following six tables present an overview of the display elements depicted and the functionality provided by the surface moving map displays included in this industry snapshot. The tables provide the following information:

>Table A-3. Summary of display element depiction: Vendor displays.
>Table A-4. Summary of display element depiction: Research displays.
>Table A-5. Summary of indicators: Vendor displays.
>Table A-6. Summary of indicators: Research displays.
>Table A-7. Summary of functionality: Vendor displays
>Table A-8. Summary of functionality: Research displays.

In each of these tables, empty cells are used to indicate those display elements, indicators, or functionality that the author did not know the exact depiction or implementation or whether it is even depicted or implemented. Display elements, indicators, and functionality that are not depicted are indicated with "---" in the table cell. Detailed information for each of these displays can be found in the sections listed in the tables.

Appendix A: Industry Review 70

Prototype Representation: Vendors

Display Element	ACSS	Diehl Avioniks	Garmin AT AT2000	Honeywell	Jeppesen JeppView	Jeppesen TPA	Rockwell-Collins, Inc.	Rockwell-Collins EFB	Smiths Avionics
Section	A.1.1	A.1.2	A.1.3.2	A.1.4	A.1.5.1	A.1.5.2	A.1.6.1	A.1.6.2	A.1.7
Status	Prototype	Prototype	TSO	Prototype	Prototype	Prototype	Prototype	Prototype	Prototype
Ownship	White aircraft icon	Yellow aircraft icon	White isosceles triangle	White stick figure aircraft icon	Gray aircraft icon, outlined in black	Blue isosceles triangle, outlined in white	Cyan isosceles triangle	White isosceles triangle, filled	White isosceles triangle
Traffic	In air: blue On ground: brown Icons represent surveillance technology transmitting the signal, e.g.: △ ADS-B ◁ TIS-B ◆ TCAS intruder ◇ TCAS traffic ● traffic advisory	In air: blue diamond On ground: yellow circle	On ground: brown In air: blue ADS-B aircraft TIS-B aircraft TCAS traffic alert TCAS resolution advisory TCAS other traffic ground vehicle	In air: blue On ground: brown Icons represent surveillance technology transmitting the signal △ ADS-B ◁ TIS-B ◆ TCAS intruder ◇ TCAS traffic ● advisory			In air: white On ground: brown chevrons ADS-B aircraft ADS-B (undetermined direction) TIS-B TCAS traffic alert TCAS resolution advisory	In air: white On ground: 2 implementations -- brown or brown, outlined in white Icons used are identical to that used for Rockwell Collins in-cockpit surface moving map	
Selected traffic	Outlined		Highlighted outlined in green				Highlighted in green. Position represented when target aircraft is outside map range	Highlighted – circular region surrounds traffic	

Table A-3. Summary of display element depiction: Vendor displays.
Display elements that are not depicted are indicated with "--". Empty cells indicate those display elements that the author did not know the depiction, if depicted at all.

Appendix A: Industry Review

Prototype Representation: Vendors

Display Element	ACSS	Diehl Avioniks	Garmin AT AT2000	Honeywell	Jeppesen JeppView	Jeppesen TPA	Rockwell-Collins, Inc.	Rockwell-Collins EFB	Smiths Avionics
Section	A.1.1	A.1.2	A.1.3.2	A.1.4	A.1.5.1	A.1.5.2	A.1.6.1	A.1.6.2	A.1.7
Runways	White	Light gray	Gray	White	Black	Light gray - filled	Light gray - filled	Light gray	Black, outlined with white dashed lines. Runway ends are marked with white circles. Runway-taxiway intersections marked with green circles.
Runway labels		White	White	Only exits labeled	Black	Cyan text in black box that is outlined in cyan	Black text in white box	Two implementations: (1) Black text on gray, white text on red (2) White text on red	White
Runway centerlines	---	White "^" symbol	---	Light yellow	---	---	Light gray or white	Yellow	---
Taxiways	black	Dark gray	Black	Gray	Gray	Dark gray	Black with blue edge lines	Black	Black, outlined with cyan dashed lines
Taxiway labels	---	Yellow	Yellow	Black text	Black	Cyan	Green	Two implementations: (1) white (2) amber	Green
Taxiway centerlines	---	---	---	Light yellow	---	---	Gray	Gray	---
Edge lines	---		White	Blue	---	---	Light yellow/ white	---	---

Table A-3. Summary of display element depiction: Vendor displays. (continued)

Display elements that are not depicted are indicated with "---". Empty cells indicate those display elements that the author did not know the depiction, if depicted at all.

Display Element	ACSS	Diehl Avioniks	Garmin AT AT2000	Honeywell	Jeppesen JeppView	Jeppesen TPA	Rockwell-Collins, Inc.	Rockwell-Collins EFB	Smiths Avionics
Section	A.1.1	A.1.2	A.1.3.2	A.1.4	A.1.5.1	A.1.5.2	A.1.6.1	A.1.6.2	A.1.7
Hold lines	---	Green, if ownship has clearance to cross runway Red, if ownship does not clearance to cross runway	Yellow	yellow	---	Yellow	Yellow	Yellow	Magenta
Hold short directives	---		---	---	---	---	---	---	Magenta text displays the hold command, e.g., (HLD 13R)
Non-movement areas		Black	Black		White	Black	Black	Black	black
Grassy areas	Bright green	Black	Green		White	Black	Black	Green	
Buildings	Bright green	Black	Blue	Blue outlines	Black	Dark blue	Black, outlined in blue	Two implementations: (1) blue (2) brown	Black
Building labels	---	Blue	---	Black text	black	Cyan	---	---	
Fence line	---	---	---	---	---	---	---	Cyan	---
Stand areas	---	---	---	---	---	---	---	Gray	---
Taxi route		Green line; parking spot at gate indicated with green circle	---		---	---	---	---	Magenta; waypoints indicated by magenta circles
ATC clearances and messages	---		---	---	---	---	---	---	

Table A-3. Summary of display element depiction: Vendor displays. (continued)

Display elements that are not depicted are indicated with "---". Empty cells indicate those display elements that the author did not know the depiction, if depicted at all.

Appendix A: Industry Review 73

Display Element	Prototype Representation: Research Organizations		
	FAA-Tech Center	**MITRE**	**NASA-Ames**
Section	A.2.1	A.2.2	A.2.3
Ownship	White isosceles triangle	White isosceles triangle	White chevron
Traffic	In air: blue On ground: brown ADS-B in air ADS-B on ground TIS-B in air TIS-B on ground TCAS traffic alert TCAS resolution advisory TCAS other traffic ground vehicle	In air: cyan chevrons On ground: brown chevrons ADS-B aircraft in air ADS-B aircraft on the ground TIS-B aircraft in air TIS-B aircraft on the ground	White aircraft icon
Selected traffic	Highlighted and outlined in green	Highlighted and outlined green chevron selected aircraft in air selected aircraft on the ground	
Runways	Light gray	Gray, may be outlined in white or filled-in	Black. Red outline indicates occupancy
Runway labels	Yellow	N/A due to database issue	White
Runway centerlines	---	---	---

Table A-4. Summary of display element depiction: Research displays.

Display elements that are not depicted are indicated with "---". Empty cells indicate those display elements that the author did not know the depiction, if depicted at all.

Display Element	Prototype Representation: Research Organizations		
	FAA-Tech Center	MITRE	NASA-Ames
Section	A.2.1	A.2.2	A.2.3
Taxiways	Dark gray	Black / lack of color	Black
Taxiway labels	Light yellow	White labels that rotate to remain oriented, autoscaled as a function of map range	white
Taxiway centerlines	---	---	---
Edge lines	White	Blue	
Hold lines	---	Yellow lines	Red bar
Hold short directives	---	---	Yellow Flashing hold bar for ownship and traffic
Non-movement areas	Gray	Black	Black
Grassy areas	Black	One of four options that were examined: light or dark green, light or dark gray. Initial research showed pilots preferred dark subtle colors (Bone, et al., 2003).	Green
Buildings	Dark gray, cross-hatched texturing	Black, outlined in beige	Blue
Building labels	---	---	---
Taxi route	---	---	Pending: white, flashing Cleared: magenta
ATC clearances and messages	---	---	Text box below map

Table A-4. Summary of display element depiction: Research displays. (continued)

Display elements that are not depicted are indicated with "---". Empty cells indicate those display elements that the author did not know the depiction, if depicted at all.

Prototype Representation: Vendors

Indicators	ACSS	Diehl Avioniks	Garmin AT AT2000	Honeywell	Jeppesen JeppView	Jeppesen TPA	Rockwell-Collins	Rockwell-Collins EFB	Smiths Avionics
Section	A.1.1	A.1.2	A.1.3.2	A.1.4	A.1.5.1	A.1.5.2	A.1.6.1	A.1.6.2	A.1.7
Lubber line	---				---			---	White
Range ring/ compass rose		White	White. Compass rose shows 330.	White. Compass rose available in track-up mode. Shows full and ½ range.		White	White. Note range ring and compass rose are separate features.	White. Note range ring and compass rose are separate features.	White
Ground track/ velocity vector			White				Magenta or white segmented line; each segment corresponds to a 20s interval		---
Background color	Black		Black	Gray	White	Black	Black	Black	Black

Table A-5. Summary of indicators: Vendor displays.

Indicators that are not depicted are indicated with "---". Empty cells indicate those indicators that the author did not know the depiction, if depicted at all.

Prototype Representation: Research Organizations

Indicators	FAA-Tech Center	MITRE	NASA-Ames
Section	A.2.1	A.2.2	A.2.3
Lubber line	White	White	---
Range ring/ compass rose	White. Shows full and ½ range	Yellow	---
Ground track/velocity vector	White	White circle, at nose of aircraft	
Background color	Black	Black	Black

Table A-6. Summary of indicators: Research displays.

Indicators that are not depicted are indicated with "---". Empty cells indicate those indicators that the author did not know the depiction, if depicted at all.

Functionality	ACSS	Diehl Avioniks	Garmin AT AT2000	Honeywell	Jeppesen FliteDeck	Jeppesen TPA	Rockwell-Collins	Rockwell-Collins EFB	Smiths Avionics
Section	A.1.1	A.1.2	A.1.3.2	A.1.4	A.1.5.1	A.1.5.2	A.1.6.1	A.1.6.2	A.1.7
Prioritization of map features			---	---	---	---	✓		
Map Range (Zoom)	✓		✓	✓		✓	✓	✓	
Autozoom			✓		---			✓	
Declutter	✓		(tied to map range)	✓	---	✓	✓	✓	
Traffic display	✓		✓	✓		(tied to map range)	(tied to map range)	✓	
Traffic selection			✓		✓		✓	✓	
Altitude filter	✓		✓	✓			✓	✓	
Panning					✓	✓	---	✓	
Toggle: North-up/ Track-up					---	✓	---	✓	
Flight ID			✓				✓		
Range ring			✓		---	✓	✓	✓	
Velocity Vector			✓		✓		✓	✓	
Compass / forward arc					---		✓	✓	

Table A-7. Summary of functionality: Vendor displays

Functionality that is not available is indicated with "---". Empty cells indicate those functions that the author did not know if it was available.

Functionality	Prototype Implementation: Research Organizations		
	FAA-Tech Center	MITRE	NASA-Ames
Section	A.2.1	A.2.2	A.2.3
Prioritization of map features			
Map Range (Zoom)	✓	✓	✓
Autozoom	✓	✓	
Declutter	✓	✓	✓
Traffic Display	✓	✓	✓
Traffic Selection	✓	✓	
Altitude Filter		✓	
Panning	---	---	---
Toggle functions:			
North-up/Track-up	✓	✓	✓
Flight ID	✓	✓	✓
Range ring	✓	✓	
Velocity Vector	✓	✓	
Compass / forward arc	✓	✓	

Table A-8. Summary of functionality: Research displays.

Functionality that is not available is indicated with "---". Empty cells indicate those functions that the author did not know if it was available.

A.1 INDUSTRY PROTOTYPES

A.1.1 ACSS (Aviation Communication & Surveillance Systems)   Phoenix, Arizona
Website: www.l-3com.com/acss/   Status: Prototype

Display elements
The following display elements are depicted on the ACSS surface moving map:

Display element	Representation
Ownship	White aircraft icon
Traffic	Aircraft on the ground color-coded brown; aircraft in air color-coded blue. Icons represent the data surveillance technology transmitting the signal: △ ADS-B ⌂ TIS-B ◆ TCAS intruder ◇ TCAS traffic ● traffic advisory ☐ alert
Selected traffic	Selected traffic is outlined
Runways	White
Taxiways	Black
Grassy areas	Bright green
Buildings	Bright green

Table A-9. ACSS display elements.

Functionality

The ACSS surface map is part of a multi-function display. Eight customization modes are provided along the left and right edges of the surface map, as shown in Figure A-1, and functionality within each mode is selectable using the five buttons located at the bottom of the display. Labels describing the functionality for each mode are printed next to the button. Text labels are oriented vertically.

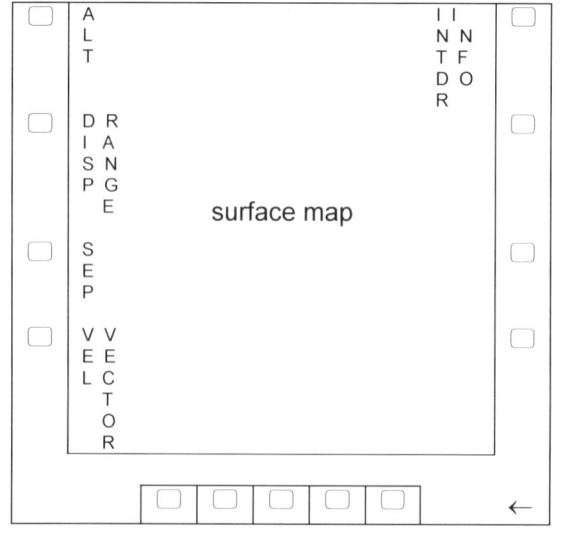

Figure A-1. Control mode for ACSS map display.

As an example of the control interface, when intruder information is selected (the top right button in Figure A-1), the user is then presented with the options for filtering TCAS targets, ADS-B targets, etc along the bottom of the display.

The ACSS display provides the following functions:
- Map range (zoom)
- Declutter
- Toggle velocity vector on/off
- Altitude filter
- Traffic display
- Traffic selection

Functions that were not demonstrated (or that are not available) include:
- Prioritization of map features
- Autozoom
- Panning
- Toggling between north-up and track-up orientations
- Toggle flight ID on/off
- Toggle range ring on/off
- Switch between compass and forward arc views

A.1.2 Diehl Avionik Systeme
Website: www.diehl.com

Frankfurt, Germany
Status: Prototype

Display elements

Diehl Avionik Systeme, in conjunction with Darmstadt University, has developed a prototype display to support airport navigation for the Airbus A380. The system is intended to support the following four functions:

- aid familiarization of the airport surface, especially at unfamiliar airports, and serve as a replacement to paper charts
- enhance airport navigation by providing situational information, destination and path instructions, and taxi information
- provide surveillance to improve position awareness while taxiing and parking
- provide guidance

Figure A-2 presents an image of the display prototype.

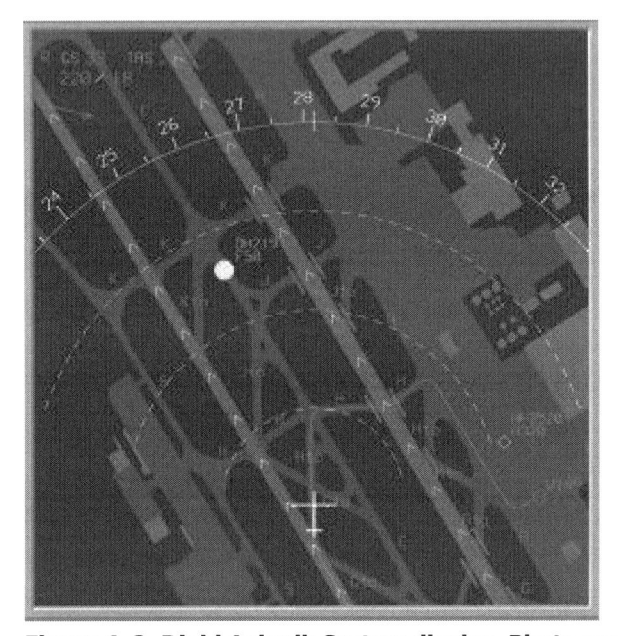

Figure A-2. Diehl Avionik System display. Photo courtesy of Diehl Avionik.

As shown in Figure A-2, the Diehl surface map is a track-up display. The following elements are depicted:

Display element	Representation
Ownship	Yellow aircraft icon
Traffic	On ground – yellow circle; In air – blue diamond
Traffic ID	Yellow
Runways	Light gray
Runway centerlines	White "^" symbol
Runway labels	White
Taxiways	Dark gray
Taxiway labels	yellow
Hold lines	Green, if ownship has clearance to cross runway Red, if ownship does not clearance to cross runway

Table A-10. Diehl Avionik display elements.

Display element	Representation
Non-movement areas	black
Grassy areas	Black
Buildings	Black
Building labels	Blue
Taxi route	Green line; parking spot at gate is indicated with a green circle

Table A-10. Diehl Avionik display elements. (continued)

Additionally, a white range ring/compass rose is provided at the top of the display to indicate ownship status.

<u>Functionality</u>

The functionality for the surface map was not provided.

A.1.3 Garmin Advanced Technologies (AT) Salem, Oregon
Website: www.garminat.com

A.1.3.1   MX20   Status: TSO (C110a, C113, C63c, C118, C147)

Display elements

The MX20 is a 6" (640 x 480 pixels) multi-function display, on which electronic versions of airport-surface diagrams may be viewed with ownship position overlaid. The MX20 airport charts are based on the JeppView product. A description of the display elements presented is described in Section A.1.5.1.

Figure A-3 presents an example of the MX20 surface display.

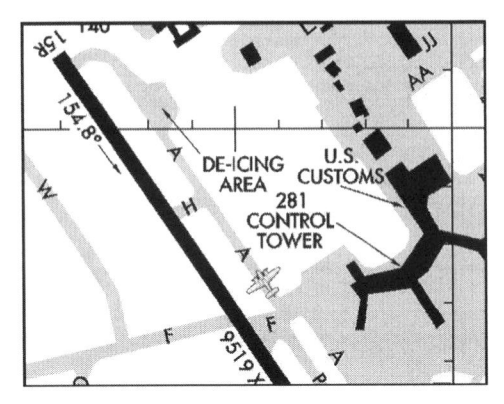

Figure A-3. MX20 display. Photo courtesy of Garmin AT.

The MX20 presents the surface map at departure, transitioning to an enroute chart after takeoff. On approaches, the MX20 presents the surface map once ownship has approached the runway and ground speed has slowed to 50 knots.

More information on the MX20 can be found at www.garminat.com/mx20_gen.shtml.

Functionality

The MX20 controls provides six "line select" keys to the right of the display and four general purpose "smart keys" at the bottom of the display, as shown in Figure A-4.

Figure A-4. MX20 controls. Photo courtesy of Garmin AT.

The Garmin AT MX20 provides the following functions:
- Map range (zoom)
- Traffic display
- Traffic selection

Functions that were not demonstrated (or that were not available) include:
- Prioritization of map features
- Altitude filter
- Autozoom
- Decluttering: implementation is tied to map range, so that the number of display elements depicted increases as the user zooms in
- Panning
- Toggling between north-up and track-up orientations
- Toggle flight ID on/off
- Toggle range ring on/off
- Toggle between compass and forward arc views

A.1.3.2 AT2000   Status: TSO (C113, C105, C119b)

Display elements

Garmin AT is also developing an advanced surface display; the AT2000 is a 6.1" multi-function CDTI that presents TCAS proximate traffic, ADS-B, and TIS-B traffic information superimposed over a track-up display. Other functions available on the AT2000 is the display of weather radar (ARINC 708 interface), terrain from TAWS and EGPWS, and flight plan and navigation information from FMS. A range ring feature is also available to aid pilots in maintaining approach and enroute spacing distances from other ADS-B traffic.

The display elements and indicators depicted on the AT2000 are listed in the tables below.

Display element	Representation
Ownship	White isosceles triangle
Traffic	Aircraft on the ground color-coded brown; aircraft in air color-coded blue. Icons represent the data surveillance technology transmitting the signal: ADS-B aircraft in air ADS-B aircraft on ground TIS-B aircraft in air TIS-B aircraft on ground TCAS traffic TCAS traffic alert TCAS resolution advisory ground vehicle
Selected traffic	Selected target in air is green and outlined in green; selected target on ground is brown and outlined in green selected aircraft in air selected aircraft on ground Information for selected target presented in green at the bottom left corner of the display

Table A-11. Garmin AT AT2000 display elements.

Display element	Representation
Runways	Gray
Runway labels	White
Taxiways	Black
Taxiway labels	Yellow
Edge lines	White
Hold lines	Yellow
Non-movement areas	Black
Grassy areas	Green
Buildings	Blue

Table A-11. Garmin AT AT2000 display elements. (continued)

Indicators	Representation
Lubber line	White
Range ring/compass rose	White. Compass rose shows 330°. Text is rotated to correspond with its position on the rose.
Ground track/velocity vector	White

Table A-12. Garmin AT AT2000 indicators.

000 provides the following functions:
- Map range (zoom)
- Autozoom
- Decluttering: implementation is tied to map range, so that the number of display elements depicted increases as the user zooms in
- Traffic display
- Traffic selection
- Toggle flight ID on/off
- Toggle range ring on/off

Functions that were not demonstrated (or that were not available) include:
- Prioritization of map features
- Panning
- Toggling between north-up and track-up orientations
- Toggle between compass and forward arc views

In the installation of the AT2000 system on Boeing 757s, a 6" x 1.5" keypad mounted on the center panel will be used to control the surface moving map, located in the center of the cockpit. The keypad layout is shown below in Figure A-5; the functionality is described in Figure A-6.

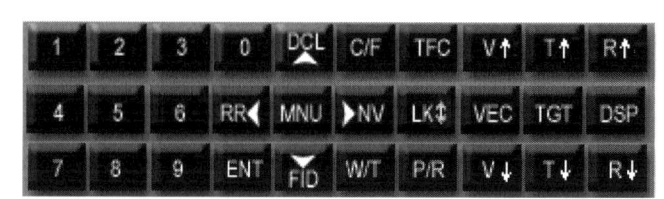

Figure A-5. Garmin AT AT2000 control layout.

	DCL Removes all display data except target icon and altitude data.	**C/F** Toggles display of Captain or FO nav data	**TFC** Toggles between TCAS and hybrid display	**V↑** Increase time value	**T↑** Highlights the next farther ADS-B target.	**R↑** Increase display range.
RR Toggles selectable range ring on/off	**MNU** Displays pages for equipment set-up. Cycle through menu with cursor keys.	**NV** Displays nav information when available.	**LK** Altitude filter: Selectable range 1500-24500, incremented by 1000'. Toggle between look-up, level, and look-down. Default values: • LVL - 2500 above and below • LK↑ - 9500 above, 2500 below • LK↓ - 2500 above, 9500 below	**VEC** Turns velocity vector on/off. Vector Time Up and Vector Time Down keys adjust time of vector display from 30sec - 6min	**TGT** Toggles target selection on/off. When selected, nearest ADS-B target is highlighted.	**DSP** Toggles between 360 compass rose and 120 arc view
ENT Accepts edited fields and returns to previous display	**FID** Toggles flight ID display on/off for all displayed targets reporting ID.	**W/T** Toggles between display of weather or terrain data, if available from aircraft system.	**P/R** Toggles target altitude between relative altitude and pressure altitude.	**V↓** Decrease time value	**T↓** Highlights the next nearer ADS-B target.	**R↓** Decrease display range.
Numeric keypad						

Figure A-6. Garmin AT control functionality.

Note: In Menu mode the DCL, NV, FID, and RR keys function as cursor keys to navigate between menu fields.

A.1.4 Honeywell

Website: www.honeywellaerospace.com

Phoenix, Arizona
Status: Prototype

Display elements

The Honeywell moving map was designed to be a subset of their Primus Epic system, which displays high-resolution terrain information with weather data. A wide range of display viewpoints is available from a close up view of the airport surface with runways and taxiways to a terrain display of the entire United States.

The display elements depicted on the Honeywell Primus Epic system are listed in Table A-13.

Display element	Representation
Ownship	White stick figure aircraft icon (e.g., ↦)
Traffic	Aircraft on the ground color-coded brown; aircraft in air color-coded blue. Icons represent the data surveillance technology transmitting the signal.
Runways	White
Runway labels*	-- Only runway exits are labeled
Runway centerlines*	Light yellow
Taxiways*	Gray
Taxiway labels*	Black text in a font size larger than that used for labeling buildings
Taxiway centerlines*	Light yellow
Edge lines*	Blue
Hold lines*	Yellow
Buildings*	Blue outlines
Building labels*	Black text

Table A-13. Honeywell display elements.

Note: The (*) indicates display elements that are not implemented on Epic displays but are depicted for prototype surface moving maps. These display elements can be displayed on Epic displays in the future, when more functional, regulatory, and certification guidance is available.

A white compass rose appears when the surface map is presented in track-up mode. Two rings overlay the surface; the outer ring is the compass rose, the inner ring shows ½ range.

Functionality

The controls used to manipulate the display are shown in Figure A-7.

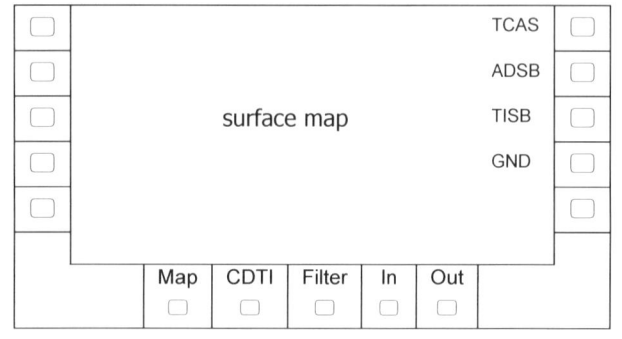

Figure A-7. Honeywell controls.

The buttons at the bottom of the control box allow for five modes: map, CDTI, filter, zoom in, and zoom out. The CDTI mode is selected in the figure. With the exception of **In** and **Out** buttons, selecting the mode changes the options at the right edge of the display.

The Honeywell display provides the following functions:
- Map range (zoom)
- Decluttering: implementation is tied to map range, so that the number of display elements depicted increases as the user zooms in. Because the Honeywell surface map is only a subset of a larger display suite (the Primus Epic system), the map range may be manipulated from a close-up view of the airport surface to a terrain display of the United States; a world map is available at the highest zoom level. At closer ranges, VOR station location information is available, followed by the presentation of ILS feathers indicating the location of runways. Taxiway labels appear when the display is zoomed to 0.9 range.
- Filter the display of traffic as a function of the technology transmitting the data

Functions that were not demonstrated (or that were not available) include:
- Prioritization of map features
- Autozoom
- Select traffic
- Panning
- Toggle between north-up and track-up orientations
- Toggle flight ID on/off
- Toggle range ring on/off
- Toggle between compass and forward arc views

A.1.5 Jeppesen
Englewood, Colorado
Website: http://www.jeppesen.com

A.1.5.1 JeppView

JeppView is a suite of applications that provide electronic aeronautical charts in various configurations for both ground-based and in flight use. Jeppview can be used on desktop, laptop and tablet computers, and also is available for select panel-mount avionics equipment. An example is shown in Figure A-8a.

(a) (b)

Photo courtesy of Garmin AT.

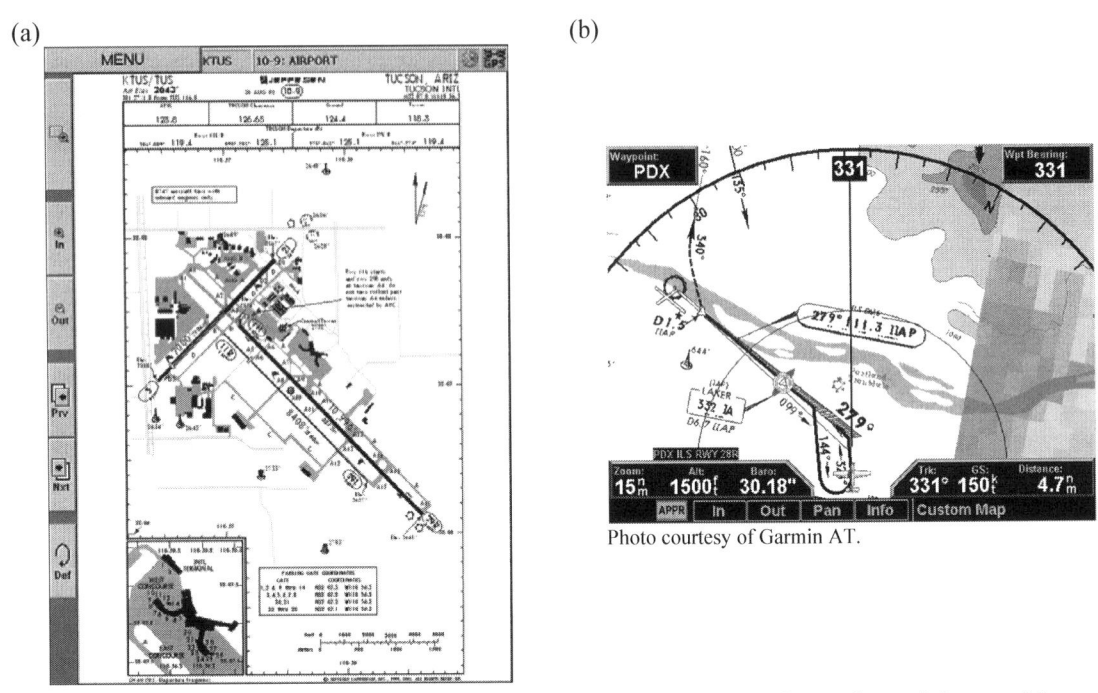

Figure A-8. (a) JeppView airport diagram, (b) Garmin AT MX-20 chart view with ownship position superimposed on a JeppView instrument approach chart.

The JeppView charts provide the same coverage as Jeppesen's Airway Manuals. Two different data sets are available: a standard data set intended for ground use only, e.g., for flight planning, and a geo-referenced data set which may be used in the air and allows the depiction of ownship position, if the aircraft is equipped with a GPS signal feed. The latitude/longitude data for each feature on geo-referenced charts is verified to ensure that ownship position is accurate with respect to the depicted airport features. This geo-referenced chart view can be transferred from a computer to a flash memory card for installation in a panel mounted display. As shown in Figure A-8b, the Garmin AT MX-20 (discussed in Section A.1.3.1) uses JeppView instrument approach charts and airport diagrams in their moving map display. In the figure, ownship is presented at the bottom of the display, superimposed on an instrument approach chart.

Display elements

The display elements presented in the JeppView airport diagram and the method of depiction is described in Table A-14.

Display element	Representation
Ownship	Gray aircraft icon outlined in black
Runways	Black
Runway labels	Black
Taxiways	Gray
Taxiway labels	Black

Table A-14. JeppView FliteDeck display elements.

Display element	Representation
Non-movement areas	White
Grassy areas	White
Buildings	Black

Table A-14. JeppView FliteDeck display elements. (continued)

Functionality

JeppView FliteDeck provides the following functions:
- Map range (zoom)
- Panning

Functions that were not demonstrated (or that were not available) include:
- Prioritization of map features
- Autozoom
- Altitude filter
- Traffic display
- Traffic selection
- Decluttering
- Toggling between north-up and track-up orientations
- Toggle flight ID on/off
- Toggle range ring on/off
- Toggle between compass and forward arc views

A.1.5.2   Airport Moving Map(AMM)

Jeppesen has developed an Airport Moving Map (AMM) application to assist operators in the improvement and efficiency of ground operations. The AMM does not provide primary guidance and is designed to supplement current regulations and operational practices for the taxi environment. An image of the AMM application is presented in Figure A-9.

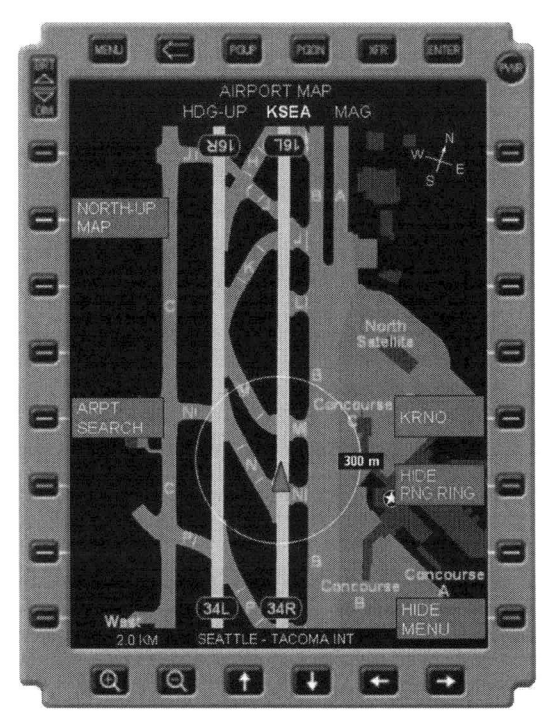

Figure A-9. Jeppesen TPA System. Photo courtesy of Jeppesen.

The AMM application has been developed and deployed on Class 3 EFB in conjunction with Boeing and Astronautics Corporation. A Class 2 version of the application will be available for a variety of additional platforms in early 2005.

Display elements

The display elements presented on the AMM application and their method of depiction is described in Table A-15.

Display element	Representation
Ownship*	The fill and outline colors of the Ownship symbol are supplier configurable items
Runways	Light gray – filled
Runway labels	White text in black box
Taxiways	Dark gray
Taxiway labels	White
Hold lines	Amber
Non-movement areas	Black
Buildings	Blue
Building labels	White

Table A-15. Jeppesen TPA display elements.

Table A-16 lists the indicators presented on the Jeppesen AMM display.

Indicators	Representation
Range ring	White circle
Compass	White

Table A-16. Jeppesen TPA indicators.

Functionality

The Jeppesen AMM application provides the following functions:
- Preset Map Ranges (zoom)
- Decluttering (tied to map range with less detail shown at higher map ranges)
- Toggle range ring on/off
- Toggle between north-up and track-up orientations
- Panning
- Ownship indicator *
    - Directional ownship when valid heading available
    - Non-directional ownship when no heading available
    - Ownship removed when ANP < RNP
- High precision airport map database

Functions that were not demonstrated (or that were not available) include:
- Altitude filter
- Traffic display
- Traffic selection

---

* At this time, it is expected that ownship position will be depicted on the Class 3 EFB but not on the Class 2 EFB.

- Prioritization of map features
- Autozoom
- Select traffic
- Toggle flight ID on/off
- Toggle between compass and forward arc views

A.1.6 Rockwell Collins, Inc. Cedar Rapids, Iowa
Website: www.rockwellcollins.com

Rockwell Collins has developed two prototypes: one to be mounted directly into the cockpit, the other to be displayed on a carry-on device, such as an EFB. The display elements depicted and functionality implemented differs between the two displays, as described below. Since these are prototype displays, they are subject to frequent updates. The following information is from May 2002. The images of the surface map display and the hardware are snapshots in time of a prototype under evaluation.

NOTE: No recommendations are made by Rockwell Collins as to the suitability of the display, display elements, or controls for use on aircraft flight decks.

A.1.6.1   Cockpit Display Status: Prototype

Display elements

The Rockwell Collins cockpit display is presented in a track-up format, superimposed over a 360° compass rose; currently, no north-up mode is available. An image of the display is presented in Figure A-10.

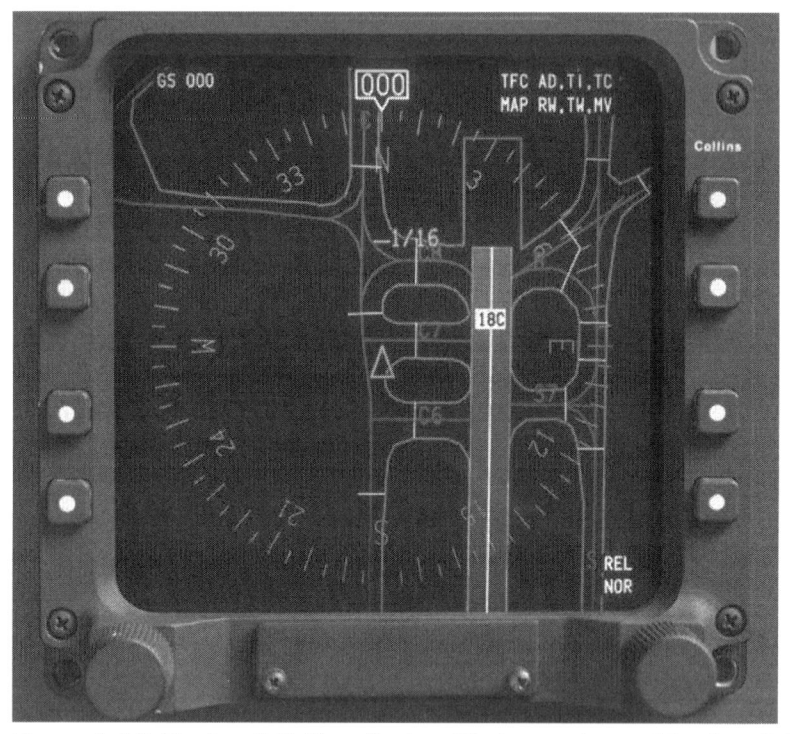

Figure A-10. Rockwell Collins display. Photo courtesy of Rockwell Collins.

The display elements presented are listed in Table A-17.

Display element	Representation
Ownship	cyan isosceles triangle
Traffic	Traffic in the air is represented in white; traffic on the ground represented in brown. The icon presented is representative of the technology transmitting the signal. The icons used were based on the specifications for the Test and Evaluation Surveillance Information System (TESIS) demonstration and are subject to change. ADS-B aircraft in the air ADS-B aircraft on ground ADS-B aircraft in the air (undetermined direction) TIS-B aircraft in the air TIS-B aircraft on the ground TCAS other traffic TCAS traffic alert TCAS resolution advisory surface vehicle
Selected traffic	The selected traffic is highlighted in green. Information regarding the target ID, category, ground speed, and range is printed in green text in the left corner of the display. When the target is selected but the position of the target aircraft is outside of the map range, a green icon appears at the edge of the compass rose to indicate the relative position of the target aircraft
Runways	Light gray filled rectangle
Runway labels	Black text in white box
Runway centerlines	Light gray or white
Taxiways	Black with blue edge lines
Taxiway labels	Green
Taxiway centerlines	Gray
Edge lines	Light yellow/white
Hold lines	Yellow
Non-movement areas	Black
Buildings	Black, outlined in blue

Table A-17. Rockwell Collins in cockpit display elements.

Table A-18 lists the indicators presented on the Rockwell Collins surface moving map.

Indicators	Representation
Lubber line	Light gray
Range ring	White
Compass rose	Gray, numbers printed in gray text
Ground track/velocity vector	White solid line, representing the "time length"

Table A-18. Rockwell Collins indicators.

Functionality

The control functionality for the prototype presented at the FAA-sponsored demonstration was modified to suit the purposes of the demonstration. It is by no means the final implementation. The control panel is shown in Figure A-11.

Figure A-11. Rockwell Collins control panel. Photo courtesy of Rockwell Collins.

Note that Rockwell Collins makes no recommendation as to its suitability as a CDTI control panel. A dedicated control panel, multi-purpose shared control panel (like the MCDU), and/or pointing device based control panel are all potential candidates for use on the aircraft.

The control panel consists of 8 dedicated buttons, four dual function buttons (rotate and push), and one rotary knob. The Rockwell Collins display provides the following functions:
- Prioritization of map features
- Map range (zoom)
- Autozoom
- Decluttering: implementation is tied to map range, so that the number of display elements depicted increases as the user zooms in. The display elements were clustered in 3 levels:
  - Base level (always present): runways. This is the only surface attribute depicted while the aircraft is in the air.
  - + taxiways, movement areas, terminal buildings, labels for runways and taxiways
  - + ILS hold areas, service areas such as gate IDs, stands, deicing areas
- Altitude filter (ABV/BLW/NOR/ALL)
- Relative/absolute altitude indications
- Traffic display
- Traffic selection, highlighting, display of data block
- Toggle flight ID on/off
- Toggle range ring on/off, adjust range ring
- Toggle between compass and forward arc views

Functions that were not demonstrated (or that were not available) include:
- Panning
- Toggling between north-up and track-up orientations

A.1.6.2 PC Platform Status: Prototype

Display elements

The Rockwell Collins surface map for a PC platform is displayed on a Fujitsu Stylistic display, oriented horizontally in landscape mode. An example of the display is presented in Figure A-12.

Figure A-12. Rockwell Collins PC-based surface moving map. Photo courtesy of Rockwell Collins.

Two different implementations of the surface moving map are being considered. The display elements depicted, and the means of representation used for the two implementations, is listed below in Table A-19. While the display elements depicted are identical across the two implementations, but the method of representation for some display elements differ. In particular, the two implementations differ in their representation of traffic on the ground (outlined or not) and in the color used to depict runway labels, taxiway labels, and buildings.

Display element	Representation
Ownship	Cyan isosceles triangle, filled in
Traffic	In air: white
	On ground: Two implementations - brown or brown with a white outline
Runways	Light gray
Runway centerlines	White
Runway labels	Two implementations: Black text on gray background or white text on red background

Table A-19. Rockwell Collins PC-based surface map display elements.

Taxiways	Black, edges are not drawn. Position of taxiways is represented with centerline location
Taxiway centerlines	Gray
Taxiway labels	Two implementations: white or amber
Hold short lines	Yellow
Non movement areas	Black
Grassy areas	Green
Buildings	Two implementations: blue or brown
Fence line	Cyan
Stand areas	Gray, lines depict detail
History dots	Green, filled circles

Table A-19. Rockwell Collins PC-based surface map display elements. (continued)

The range ring and compass rose are depicted as separate functions. The range ring is depicted in white and can be toggled on/off. The compass rose is depicted in white.

Functionality

The controls implemented in the EFB display is shown above in Figure A-12. Interaction with the buttons on the EFB display occurs via a touch screen. The controls consist of 15 buttons, as described in Table A-20.

Function	Control label	Implementation
Adjust map range	Range + Range −	Ranges between 0.125nm and 400nm. Range controls scale of the display.
Range ring	Range ring	Toggles range ring on/off
Adjust range ring	Increase Decrease	Adjusts size of the range ring. Range selectable as follows: 0.1nm – 10nm steps of 0.1nm 10nm-100nm steps of 1nm 100nm-500nm steps of 5nm
Autozoom	AZOOM	Available for takeoff and landing. On final approach, auto zoom automatically decreases map range as ownship approaches the airport. Engaged once ownship is within 20nm of the airport; "AZ" is displayed at bottom right corner of the display. Initially, the airport is depicted as a skeleton diagram, consisting only of runways. Once ownship is within 2 nm of the airport, the map switches to the 2nm range scale and shows the full airport depiction. Hold short bars are available for intersecting runways. The map changes to ½ nm range scale once ownship ground speed has dropped below 50 knots; at this time, all ground traffic is depicted.
Select traffic	Traffic Selection	Enables target selection, and selects nearest displayed target
Change traffic selection	Next, Prev	Scroll through ADS-B and TIS-B targets

Table A-20. Rockwell Collins PC-based display functionality.

Function	Control label	Implementation
Change altitude filter	Altitude filter Annunciation: NOR, ABV, BLW, ALL	Filter targets based on vertical position relative to ownship altitude. There are four settings: Normal: ± 2700 ft Above: +9900, -2700 ft Below: +2700, -9900 ft All: +9900, -9900 ft
Velocity vector	Velocity vector	Toggle velocity vector on/off
Adjust velocity vector	Increase, Decrease	Changes the selection of the velocity vector "length" in time
Declutter		Turns taxiway labels, airport buildings, and hold lines on/off
Toggle between arc and compass views	Align	Center on ownship to ownship at bottom of the display

Table A-20. Rockwell Collins PC-based display functionality. (continued)

Functions that were not demonstrated (or that were not available) include:
- Prioritization of map features
- Panning

A.1.7 Smiths Aerospace
Website: www.smithsind-aerospace.com

Grand Rapids, Michigan
Status: Prototype

Display elements

Smiths Aerospace is developing the Flight Management Computer System to provide an FMS approach to providing situation awareness on a navigation display. This system is being designed for Boeing 737-3/4/500, 6/7/8/900, and Boeing Business Jets. The Smiths moving map augments the airborne Flight Management System (FMS) software, providing ownship positional and navigational information to the B737 Head-up Guidance System and Electronic Flight Instrument System (EFIS). The prototype display is a track-up map, as shown in Figure A-13. Rudimentary stick drawings are used to demonstrate how simple diagrams can support situation awareness using common symbology.

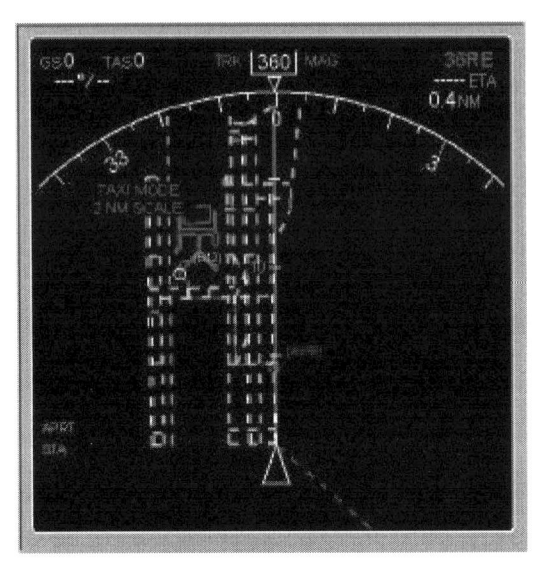

Figure A-13. Smiths Aerospace Flight Management Computer System taxi plan. Photo courtesy of Smiths Aerospace.

The viewable map range is presented in the top left quadrant of the display (in Figure A-13, the total map range shown is 2 nm). Half range is labeled in the center of the display. The taxi plan shows the terminal buildings and the runway. At take-off, the display transitions to an airborne display.

The displays will be retrofit to display on Boeing 737 aircraft. Dashed lines are used to represent runways (white) and taxiways (green), due to constraints in the display hardware, which prevent the presentation of solid lines. The lines used to represent runways and taxiways do not correspond to the actual width of the runway.

The FMCS can display the intended taxi route, if the crew enters the taxi plan. The current implementation requires that the pilot enters coordinates spaced 75 feet apart.

The display elements depicted by the FMCS are listed in Table A-21.

Display element	Representation
Ownship	White isosceles triangle
Runways	White dashed lines; runway ends are marked with white circles. Runway lines are not scaled to runway width
Runway labels	White

Table A-21. Smiths Avionics display elements.

Display element	Representation
Taxiways	Cyan dashed lines
Taxiway labels	Green
Runway/taxiway Intersections	Green circles
Hold lines	Magenta, with magenta text presented to redundantly display the hold command, e.g., (HLD 13R)
Non-movement areas	Black
Grassy areas	Black
Taxi route	Magenta; waypoints are indicated by a magenta circle

Table A-21. Smiths Avionics display elements. (continued)

The display also presents a white lubber line and white range ring/compass rose.

Functionality

A list of functions was not available. The FMCS is designed to interface with various display systems; the functions available will be associated with the display system operation.

A.2 RESEARCH PROTOTYPES

A.2.1 William J. Hughes Technical Center (FAA)
Website: www.tc.faa.gov

Atlantic City, New Jersey
Status: Research only

Display elements

The William J. Hughes Technical Center has focused its efforts towards EFB displays. The surface map is a display developed for research purposes only. It is not intended to be approved by the FAA, nor will it be installed on any aircraft.

A track-up version of the surface map is shown in Figure A-14.

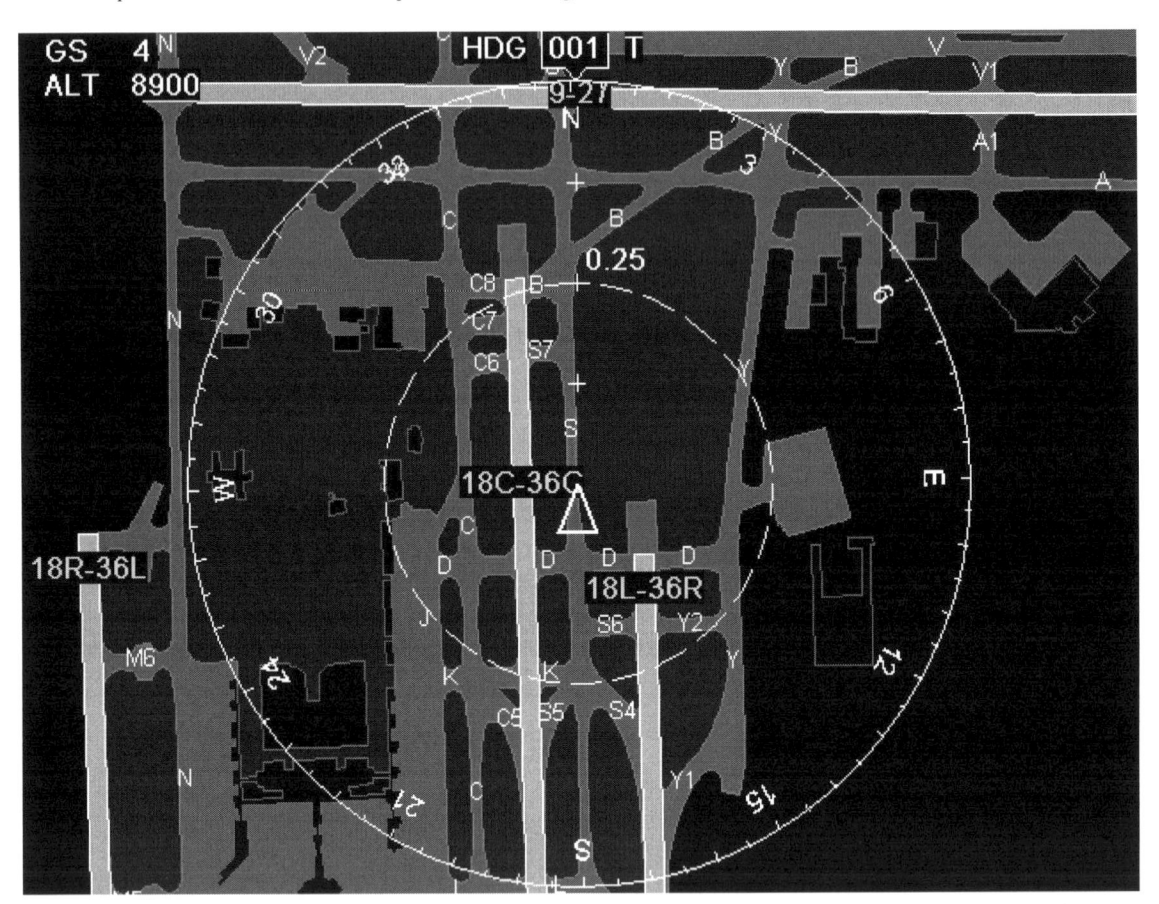

Figure A-14. FAA Tech Center surface map. Photo courtesy of William J. Hughes Technical Center.

The map presented in Figure A-14 could be presented on one of two displays: an 8.4" Fujitsu display or a 3" IPAQ hand-held display. As shown in the figure above, ownship position, represented by a white isosceles triangle, is located in the center of the display. Ownship information is presented at the top of the display, with ground speed at the left, and heading in the middle. Surrounding ownship are two range rings: one at full range, on which compass rose information is presented; the second indicates half range.

The display elements depicted and their representation are listed in Table A-22. Note that the colors and representations used are continuously being modified in order to determine an optimal coding convention. The colors listed below describe the color scheme depicted in the Figure A-14 above.

Display element	Representation
Ownship	White isosceles triangle
Traffic	Aircraft on the ground is color-coded brown; aircraft in the air is color-coded cyan. Icons represent the data surveillance technology transmitting the signal. A small subset of the symbols is presented below. ADS-B in air ADS-B on ground TIS-B in air TIS-B on ground TCAS traffic alert TCAS resolution advisory ground vehicle
Selected traffic	Selected traffic is highlighted with a green outline.
Runways	Light gray
Runway labels	Light cyan
Taxiways	Dark gray
Taxiway labels	White
Edge lines	White
Non-movement areas	Black
Grassy areas	Green
Buildings	Brown

Table A-22. FAA Tech Center display elements.

The following indicators are presented.

Indicators	Representation
Lubber line	White – can be represented by white crosses
Range ring/compass rose	White. Two range rings are displayed; one at full range, on which compass rose information is presented, the other at ½ range, with the ½ range labeled.
Ground track/velocity vector	White

Table A-23. FAA Tech Center: Indicators.

Functionality

The FAA Tech Center surface map provides the following functions:
- Map range (zoom)
- Autozoom
- Decluttering
- Traffic display
- Traffic selection
- Toggling between north-up and track-up orientations
- Toggle flight ID on/off
- Toggle range ring on/off
- Toggle velocity vector on/off
- Toggle between compass and forward arc views

Functions that were not demonstrated (or that were not available) include:
- Altitude filter
- Prioritization of map features
- Panning

Users can interact with the display via a touch-screen interface using buttons located to the right of the surface display, or with keyboard shortcut keys. Figure A-15 presents the button layout, description of the button functionality, and the corresponding shortcut keyboard command.

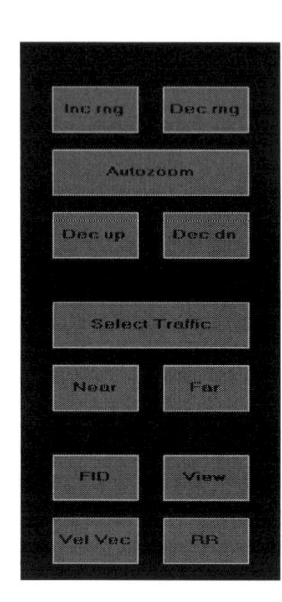

Control	Keyboard Command	Functionality
Inc rng	R	Increase map range. Loops at maximum range.
Dec rng	r	Decreases map range. Loops at minimum range
Autozoom		Toggles autozoom functionality on/off. Autozoom functionality implementation is identical to MITRE and Garmin AT implementation.
Dec up	M	Decluttering functionality: adds features. Cycles through map display modes: runways; +movement areas, non-movement areas, buildings; +runway labels, +taxiway labels. Loops at maximum level.
Dec down	m	Decluttering functionality: removes features. Loops at minimum level.
Select traffic	S/s	Toggle target selection mode. Selects nearest target or cancels selection.
Near	X/x	If target selection mode is on, then selects next closest target.
Far	W/w	If target selection mode is on, then selects next farthest target.
FID	F/f	Toggle flight ID display on/off.
View	C/c	Toggle between 360° compass mode and forward arc mode.
Velocity vector	T/t	Toggle velocity vector for ownship and traffic on/off.
RR	E/e	Toggle range ring on/off.

Figure A-15. FAA Tech Center: controls and functionality.

A.2.2 MITRE

Website: www.caasd.org

McLean, Virginia
Status: Research only

Display elements

The MITRE prototype surface display was developed for general traffic / situation awareness, using data from NACO/NGS databases. The surface moving map is presented on a navigation display, similar to a Boeing 777 or 747-400 displays.

The design of the prototype display was based on features available through airport surveys, task analyses of information needed to maintain airport surface situation awareness and final approach and runway occupancy awareness, and previous research at NASA-Ames. The map display is shown in track-up mode in Figure A-16.

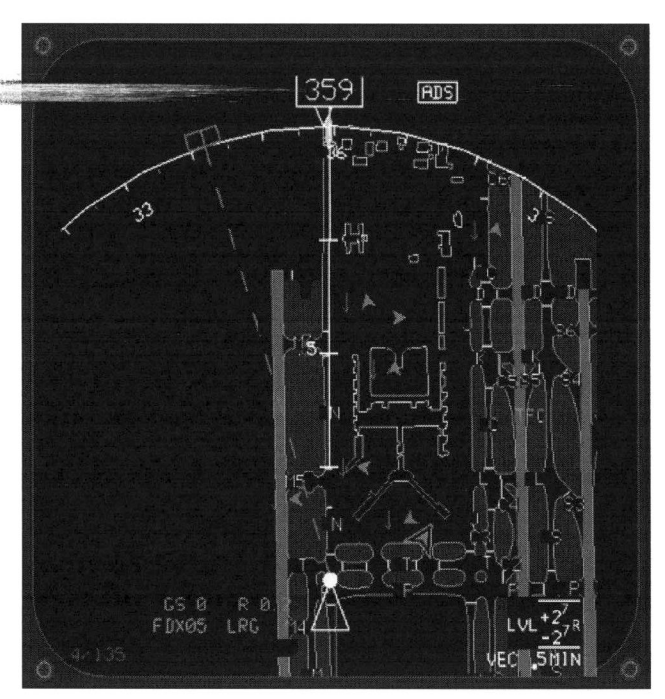

Figure A-16. MITRE surface display. Photo courtesy of MITRE.

The MITRE panel-mounted display is approximately 8" diagonal. The surface map is a track-up display. As shown in Figure A-16, ownship is represented by the white isosceles triangle and is located at the bottom of the display. Ownship information is available at the top of the display, with ground speed presented in the top left corner and digital heading in the middle. In the figure, a reduced compass rose, showing an arc of approximately 90°, redundantly provides a qualitative indication of ownship heading, but a full compass rose mode, with ownship centered on the display, is also available. A lubber line extends from the middle of the rose down, approximately 2/3 of the way down the display.

The brown chevrons in Figure A-16 show the position of traffic on the ground (Note: No airborne targets existed when this screen was captured. Therefore none are shown.). Traffic information is filtered according to ownship position with respect to the range selected and altitude band shown in the lower right corner of the display; in Figure A-16, this range is ±2700 feet of ownship altitude. Ground targets are shown on the display when ownship landing gear is down and its radar altitude is 1500 feet or below. The highlighted brown chevron shows a selected target. If the selected target were airborne, it would be shown as an outlined green chevron. Information regarding the selected target's ground speed, range from ownship, aircraft identification / call sign, and aircraft size classification is shown in the lower left corner of the display.

The display elements available and their representation are listed in Table A-24. The colors and representations used are being examined in order to determine an optimal coding convention.

Display element	Representation
Ownship	White isosceles triangle (same as that used when airborne) with the option of a closure indicator and / or vector line
Traffic	Traffic in air is represented with cyan chevrons; traffic on the ground represented by brown chevrons. A small subset of the symbols is presented below. ADS-B aircraft in air ADS-B aircraft on the ground TIS-B aircraft in air TIS-B aircraft on the ground
Selected traffic	For selected targets that are in the air, the aircraft icon is represented with a green chevron, which is outlined in green. For selected targets on the ground, the aircraft icon is a brown chevron outlined in brown. selected aircraft in air selected aircraft on the ground Aircraft information for the selected target appears in the lower left corner of the display and includes aircraft ground speed, range from ownship, flight identification / call sign, and aircraft size classification. The text is color coded, so that it is green if aircraft is airborne and brown if the aircraft is on the ground (thus "linking" visually the target and the information).
Runways	Gray, could be outlined or filled in
Runway labels	N/A due to database issue
Taxiways	Black / lack of color
Taxiway labels	White labels that rotate so they remain oriented and autoscaled depending on map range (i.e., letters increase in size as the user zooms in)
Taxiway edge lines	Blue
Hold lines	Yellow lines whose length is equal to the taxiway's width
Grassy areas	One of four color options that were examined: light or dark green, light or dark gray. Initial research indicated that pilots preferred dark subtle colors (Bone, et al., 2003)
Non-movement areas	Black / lack of color
Buildings	Black / lack of color, outlined in beige

Table A-24. MITRE surface moving map elements.

Note that although the data for taxiway centerlines were available through, they were not presented as it was determined that the presentation of taxiway centerlines did not support surface operations (Bone, et al., 2003).

The following indicators are available.

Indicators	Representation
Lubber line	White
Range ring/compass rose	Yellow
Ground track/Velocity Vector	White circle appearing at nose of aircraft

Table A-25. MITRE indicators.

Functionality

The controls for the MITRE surface moving map were adapted from the controls for the Garmin AT AT2000. See Section A.1.3.2 for more information.

The MITRE surface moving map provides the following functions:
- Map range (zoom)
- Autozoom
- Decluttering
- Traffic display
- Traffic selection
- Toggling between north-up and track-up orientations
- Toggle flight ID on/off for all aircraft
- Toggle range ring on/off
- Toggle velocity vector on/off & selectable time options
- Toggle between compass and arc mode
- Altitude filter

Functions that were not demonstrated (or that were not available) include:
- Panning

The options for the map range of the display was from 0.1 nm to 320 nm. The MITRE surface moving map was designed to provide an autozoom feature to support the final approach and runway occupancy awareness applcation. When this feature was enabled, the surface moving map would appear on the navigation display. As ownship approached the airport, the navigation display would automatically decrease in range in 0.1 mile increments, continually zooming in to a larger view of the airport surface. Map range could also be customized manually.

The surface moving map allowed pilots to selectively display different display elements, e.g., runway edge lines, taxiway edge lines, buildings, grassy areas, taxiway labels, and hold short lines. When the full airport surface was chosen to be displayed, the map depicted runways, taxiways, taxiway labels, hold short lines, grassy areas, and buildings. Note that taxiway centerlines were not an option for pilot selection; pilot interviews conducted by MITRE indicated that the presentation of centerlines resulted in display clutter. Additionally, it was believe by the researchers that presentation of taxiway centerlines did not support the airport surface situation awareness and final approach and runway occupancy awareness.

A.2.3 NASA - Ames
Website: human-factors.arc.nasa.gov/ihi/hcsl/T-NASA.html

Moffet Field, California
Status: Research only

Display elements

Researchers at NASA-Ames have developed a suite of displays to aid pilots in low visibility conditions on the airport surface; these displays together comprise the Taxiway Navigation and Situation Awareness (T-NASA) System. One of the components of the system is a surface moving map, shown in Figure A-17, which provides pilots with a track-up perspective view of the airport. A north-up overview mode, intended mainly for route planning purposes, and a Taxi ATIS mode are also available, both accessed by manipulating the range level, shown in the lower right corner of the figure.

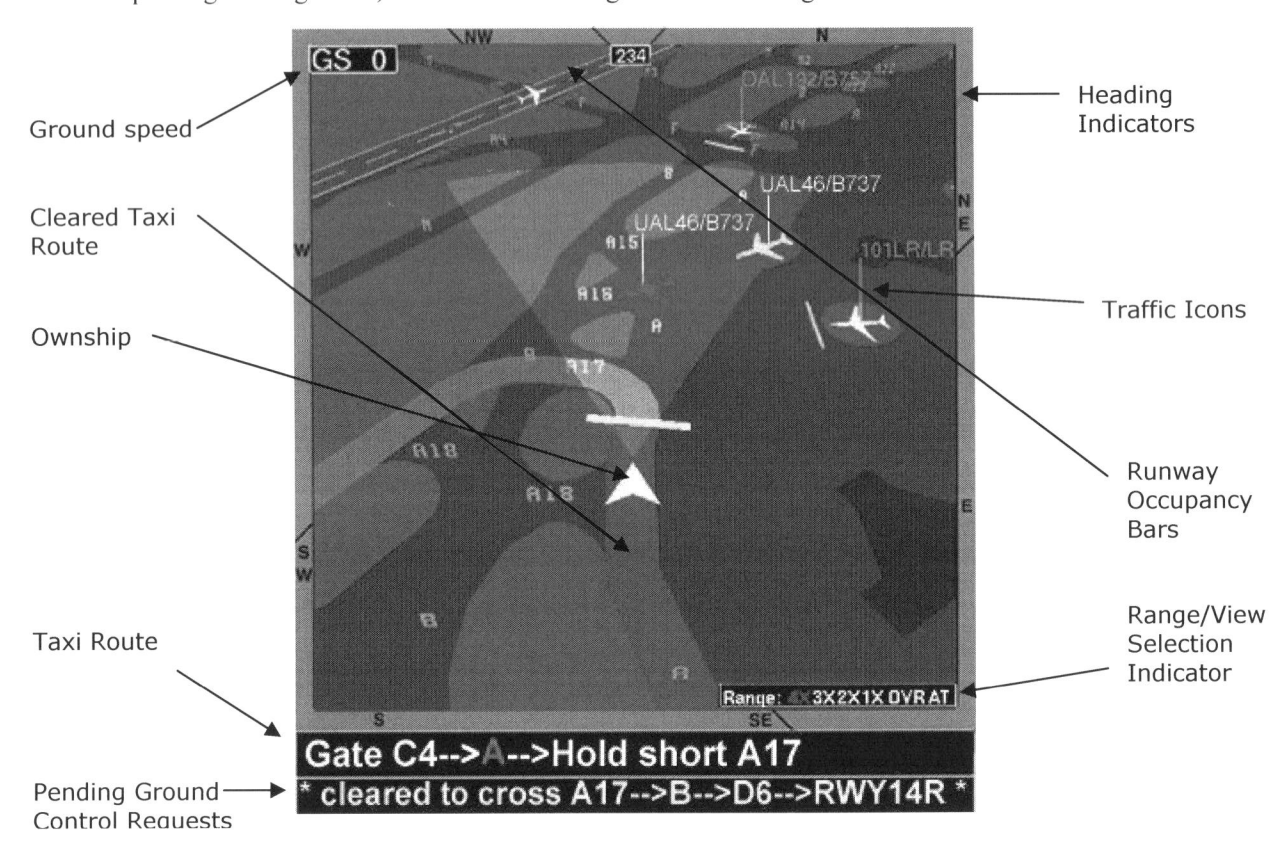

Figure A-17. T-NASA surface display. Photo courtesy of NASA-Ames.

Ownship is represented by a white triangular symbol, positioned 2/3 of the way down from the top of the display. A wedge extending from the nose of the ownship icon highlights the area directly viewable from the cockpit window and most relevant at any time to navigation and incursion information. The T-NASA display also presents ground speed and compass heading, airport traffic with ID, hold bars, route guidance, directives, and messages from air traffic control. This head-down display was integrated with a head-up counterpart that presented airborne and landing symbology to the captain during the taxi phase of flight.

It is important to note that some aspects of the design of this research display, in its current form, may preclude certification, based on guidance in regulatory documents. For example, the color red is used inappropriately in Figure A-17 to highlight the current zoom level. Regulatory documents (e.g., 14 CFR §§ 23.1322, 25.1322, 27.1322, 29.1322 and RTCA DO-257A) limit the use of red for indicating a hazard that may require immediate corrective action. Additionally, although efforts to increases runway occupancy awareness and reduce runway incursions could be considered positive features, as implemented here, the runway occupancy bars would continuously flash on/off, a condition that may be too distracting, making it difficult for the display to comply with 14 CFR § 23.1523.

The display elements depicted on the T-NASA display are presented below in Table A-26.

Display element	Representation
Ownship	White triangular symbol
Traffic	White aircraft icon, when accurate directional information is available. Otherwise, a solid circle is used with optional ID (aircraft type and call sign) Three stage TCAS color-coding scheme used to indicate potential incursions.
Runways	Black. Red lines across intersections indicate occupancy.
Taxiways	Black
Runway/taxiway labels	White
Hold lines	Red bar surrounded by yellow border
Ramp areas	Black
Grassy areas	Green
Buildings	Blue
Taxi route	Flashing white for pending routes; magenta for cleared routes; yellow for cleared routes pending hold.
Ground Speed	White on black numbers
Compass Heading	White on black numbers
Wedge	Translucent cone-shaped area

Table A-26. T-NASA display elements.

Functionality

The T-NASA display provides the following functionality:
- Map range (zoom) over 4 levels
- North-up overview mode
- Taxi ATIS mode
- Decluttering (show traffic only within wedge zone)
- Toggle flight ID on/off
- Traffic display

Functions that were not demonstrated (or that were not available) include:
- Autozoom
- Prioritization of map features
- Panning
- Traffic selection
- Toggle range ring on/off
- Toggle velocity vector on/off
- Toggle between compass and forward arc views

For two crew simulations using a B-757, the T-NASA display shares space with the left and right side navigation displays. This is shown in Figure A-18 below.

Figure A-18. T-NASA display location for two crew simulations. Photo courtesy of NASA-Ames.

While in the air, toggling functionality is provided so that pilots can switch between the surface runway preview map and the navigation display.

Once the plane touches down, the surface moving map automatically replaces the in-flight displays. The Pilot Input Device is used to change map range, toggle between overview, ATIS and perspective map modes, and toggle the declutter function.

APPENDIX B: GUIDANCE SUMMARY

This appendix contains two sections; section B.1 is a summary of the requirements and recommendations in the document, and section B.2 provides excerpts from the Code of Federal Regulations (Parts 23, 25, 27, and 29) referenced in this document.

B.1 SURFACE MAP CHECKLIST

Use

This checklist is intended for the evaluation of surface map displays. It contains the requirements and recommendations from *Human Factors Considerations in the Design of Surface Map Displays*. Notes that provide additional guidance on interpreting the requirements from the body of the document are not included here.

The section and topic where more information can be found are cross-referenced with the corresponding sections in the document. Equipment requirements are designated with a ❖. Equipment recommendations are designated using a ❑.

2 General

2.1 Use of Color

- ❖ The accepted practice for the use of red and amber is consistent with 14 CFRs 23.1322, 25.1322, 27.1322, and 29.1322 as follows: [14 CFR §§ 23.1322, 25.1322, 27.1322, 29.1322; TSO C-165/RTCA DO-257A, 2.1.6; Chandra, et al. (2003), 2.4.8]
    - (a) Red shall be used only for indicating a hazard that may require immediate corrective action.
    - (b) Amber shall be used only for indicating the possible need for future corrective action.
    - (c) Any other color may be used for aspects not described in items a-b of this section, providing the color differs sufficiently from the colors prescribed in these items to avoid possible confusion.
- ❖ Color-coded information should be accompanied by another distinguishing characteristic such as shape, location, or text. [AC 23.1311-1A; TSO C-165/RTCA DO-257A, 2.1.6]
- ❖ No more than six colors should be used for color-coding on the map display. [TSO C-165/RTCA DO-257A, 2.1.6; SAE ARP 4032; Chandra, et al. (2003), 2.4.3]
- ❖ The colors available from a symbol generator/display unit combination should be carefully selected on the basis of their chrominance separation. Research studies indicate that regions of relatively high color confusion exist between red and magenta, magenta and purple, cyan and green, and yellow and orange (amber). Colors should track with brightness so that chrominance and relative chrominance separation are maintained as much as possible over day/night operation. Requiring the flightcrew to discriminate between shades of the same color for symbol meaning in one display is not recommended. [AC 25-11, 5.a(5); Chandra, et al. (2003), 2.4.3]
- ❑ Colors on the display should be discriminable by the typical user under the variety of lighting conditions expected in a flight deck from a nominal reference design eye point. [Chandra, et al. (2003). 2.4.3]
- ❑ Each color used in a color-coding scheme should be associated with only one meaning. [Chandra, et al. (2003), 2.4.3]
- ❑ Color-coding schemes should not conflict with flight deck standards for that particular aircraft. [Chandra, et al. (2003), 2.4.3]
- ❑ Pure blue should not be used for small symbols, text, fine lines, or as a background color. Blue is a short wavelength color. On a display containing several colors, when blue and other short wavelength colors are in focus, all other colors at long wavelengths are out of focus, and vice versa. [Cardosi and Hannon, 1999; Chandra, et al. (2003), 2.4.3]

2.2 Alerts and Reminders

- ❖ Warning information must be provided to alert the crew to unsafe system operating conditions, and to enable them to take appropriate corrective action. Systems, controls, and associated monitoring and warning means must be designed to minimize crew errors which could create additional hazards. [14 CFR §§ 23.1309(b)(3), 25.1309(b)(3)]
- ❖ If a visual indicator is provided to indicate malfunction of an instrument, it must be effective under all probable cockpit lighting conditions. [14 CFR §§ 23.1321(e), 25.1321(e), 27.1321(d), 29.1321(g)]
- ❖ Short term flashing symbols (approximately 10 seconds or flash until acknowledge) are effective attention getters. A permanent or long term flashing symbol that is noncancellable should not be used. [AC 25-11, 5.g(1)]
- ❖ Messages should be prioritized and the message prioritization scheme should be documented and evaluated. [AC 120-76A, 10.d(1) and 10.d (2); Chandra, et al. (2003), 2.4.8]
- ❑ Any use of alerts should be assessed in terms of ease of interpretation, confusion with other alerts, and for consistency with flight deck alerting philosophy.

2.3 Accessibility of Controls

- ❖ Each cockpit control must be located to provide convenient operation and to prevent confusion and inadvertent operation. [14 CFR §§ 25.777(a), 27.777(a), 29.777(a); TSO C-165/RTCA DO-257A, 2.1.5.1]
  Related Policy: 14 CFR § 23.777(a) is worded slightly differently.
- ❖ The use of controls should not cause inadvertent activation of adjacent controls. [TSO C-165/RTCA DO-257A, 2.1.5.1]
- ❖ The controls must be located and arranged, with respect to the pilot's seats, so that there is full and unrestricted movement of each control without interference from the cockpit structure or the clothing of the minimum flight crew when any member of this flight crew, from 5'2" to 6'3" in height, is seated with the seat belt and shoulder harness fastened. [14 CFR § 25.777(c)]
  Related Policy: 14 CFR §§ 23.777(b), § 27.777(b) and 29.777(b) are slightly different.
- ❖ Each flight, navigation, and powerplant instrument for use by any pilot must be plainly visible to him from his station with the minimum practicable deviation from his normal position and line of vision when he is looking forward along the flight path. [14 CFR §§ 25.1321(a), 29.1321(a)]
  Related Policy: 14 CFR §§ 23.1321(a) and 27.1321(a) are worded slightly differently.
- ❖ Controls that are normally operated by the flight crew shall be readily accessible. [TSO C-165/RTCA DO-257A, 2.1.5.2]

2.4 Design of controls

- ❖ The equipment must allow each flight crew member to perform their duties without unreasonable concentration or fatigue. [14 CFR § 25.771(a)]
  Related Policy: 14 CFR §§ 23.771(a), 27.771(a), and 29.771(a) are worded slightly differently.
- ❖ Each cockpit control, other than primary flight controls and controls whose function is obvious, must be plainly marked as to its function and method of operation. [14 CFR §§ 25.1555(a), 27.1555(a), 29.1555(a); Chandra, et al. (2003), 2.5.2]
  Related Policy: 14 CFR § 23.1555(a) is worded slightly differently.
- ❖ Each item of installed equipment must be labeled as to its identification, function, or operating limitations, or any applicable combination of these factors. [14 CFR §§ 23.1301(b); 25.1301(b); 27.1301(b); 29.1301(b); TSO C-165/RTCA DO-257A, 2.1.5.1]
- ❖ If a control can be used for multiple functions, the current function shall be indicated either on the display or on the control. [TSO C-165/RTCA DO-257A, 2.1.5.1]
- ❖ Line select function keys should acceptably align with adjacent text. [TSO C-165/RTCA DO-257A, 2.1.5.2]

- For each instrument, each instrument marking must be clearly visible to the appropriate crewmember. [14 CFR § 25.1543(b)]

  Related Policy: 14 CFR §§ 23.1543(b), 27.1543(b), and 29.1543(b) are worded slightly differently.

- The instrument lights must provide sufficient illumination to make each instrument, switch and other device necessary for safe operation easily readable unless sufficient illumination is available from another source. [14 CFR § 25.1381(a)(1)]

  Related Policy: 14 CFR §§ 23.1381(a)(1), 27.1381(a)(1), and 29.1381(a)(1) are worded slightly differently.

- The equipment shall be designed so that controls intended for use during flight cannot be operated in any position, combination or sequence that would result in a condition detrimental to the equipment or operation of the aircraft. [TSO C-165/RTCA DO-257A, 2.1.5.1]

- Controls shall provide feedback when operated. [TSO C-165/RTCA DO-257A, 2.1.5.1]

- Control operation should allow sequential use without unwanted multiple entries. [TSO C-165/RTCA DO-257A, 2.1.5.1]

- Manual controls used in flight shall be operable with one hand. [TSO C-165/RTCA DO-257A, 2.1.5.1]

- Activation or use of a control should not require simultaneous use of two or more controls in flight (e.g., pushing two buttons at once). [TSO C-165/RTCA DO-257A, 2.1.5.1]

- Controls should be designed for nighttime usability (e.g., illuminated). [TSO C-165/RTCA DO-257A, 2.1.5.1]

- Each pilot compartment must be arranged to give the pilots a sufficiently extensive, clear, and undistorted view, to enable them to safely perform any maneuvers within the operating limitations of the airplane, including taxiing takeoff, approach, and landing. [14 CFR § 25.773(a)(1)]

  Related Policy: 14 CFR §§ 23.773(a), 27.773(a)(1) and 29.773(a)(1) are worded slightly differently.

- Each pilot compartment must be free of glare and reflection that could interfere with the normal duties of the minimum flight crew. [14 CFR § 25.773(a)(2)]

  Related Policy: 14 CFR §§ 23.773(a)(2), 27.773(a)(1), 29.773(a)(2) are worded slightly differently.

- Letter keys on a keypad should be arranged alphabetically or in a QWERTY format. [TSO C-165/RTCA DO-257A, 2.1.5.3]

- If a separate numeric keypad is used, the keys should be arranged in order in a row or in a 3X3 matrix with the zero at the bottom. [TSO C-165/RTCA DO-257A, 2.1.5.3]

- If non-alphanumeric special characters or functions are used, dedicated keys should be provided (e.g., space, slash (/), change sign key (+/-), "clear" and "delete," etc.). [TSO C-165/RTCA DO-257A, 2.1.5.3]

- Where knob rotation is used to control cursor movement, sequence through lists, or cause quantitative changes, the results of such rotation should be consistent with established behavior stereotypes (Reference Sanders & McCormick, 1987) as follows: [TSO C-165/RTCA DO-257A, 2.1.5.4]

  a) For X-Y cursor control (e.g., moving a pointer across the surface of the map):
  - Knob below or to the right of the display area: clockwise movement of the knob moves the cursor up or to the right.
  - Knob above the display area: clockwise rotation of knob moves cursor up or to the left.
  - Knob to left of display area: clockwise rotation of knob moves cursor down or to the right.

  b) For quantitative displays, clockwise rotation increases values.

  c) For alphabet character selection or alphabetized lists, clockwise rotation sequences forward.

- Concentric knob assemblies should be limited to no more than two knobs per assembly. [TSO C-165/RTCA DO-257A, 2.1.5.4]

- The shape of the control should be unique and, where possible, meaningful so it can be identified directly with the function.
- Soft function keys that are inactive should either not be labeled, or use some kind of display convention to indicate that the function is not available. [Chandra, et al. (2003), 2.5.2]
- Soft function keys are typically used as multi-function keys to select one of several available functions. When the same type of function is accessed from different points in the software, the common function should appear on the same physical function key whenever possible (e.g., top right). [Chandra, et al. (2003), 2.5.2]

2.5 Design of Labels
- Labels shall be used to identify fixes, other symbols, and other information, depicted on the display, where appropriate. [TSO C-165/RTCA DO-257A, 2.2.2]
- The spatial relationships between labels and the objects that they reference should be clear, logical, and, where possible, consistent. [TSO C-165/RTCA DO-257A, 2.2.2; Chandra, et al. (2003), 2.5.2]
- Alphanumeric fonts should be simple and without extraneous details (e.g., sans serif) to facilitate readability. [TSO C-165/RTCA DO-257A, 2.2.2]
- Fix labels shall be oriented to facilitate readability. [TSO C-165/RTCA DO-257A, 2.2.2]
- Label terminology and abbreviations used for describing control functions and identifying display controls should be consistent with ICAO 8400/5 (a subset of which is included in RTCA DO-257A, Appendix A). [TSO C-165/RTCA DO-257A, 2.2.2]
- All labels shall be readable at a viewing distance of 30 inches under the full range of normally expected flight deck illumination conditions (Reference MIL STD 1472D and SAE AIR1093). [TSO C-165/RTCA DO-257A, 2.2.2]
- Soft function key labels should be drawn in a reserved space outside of the main content area. [Chandra, et al. (2003), 2.5.2]
- Labels used to identify the action associated with a soft function key should be clear to the user and brief. [Chandra, et al. (2003), 2.5.2]
- Lines should be used to connect soft labels to the control buttons they identify to minimize parallax issues.

2.6 Control layout
- Controls should be organized in logical groups according to function and frequency of use. [TSO C-165/RTCA DO-257A, 2.1.5.2]
- Controls most often used together should be located together. [TSO C-165/RTCA DO-257A, 2.1.5.2]
- Controls used most frequently should be the most accessible. [TSO C-165/RTCA DO-257A, 2.1.5.2]
- Dedicated controls should be used for frequently used functions. [TSO C-165/RTCA DO-257A, 2.1.5.2]

2.7 Presentation of Text Information
- The typeface size should be appropriate for the viewing conditions (e.g., viewing distance and lighting conditions) and the criticality of the text. [Chandra, et al. (2003), 2.4.11]
- Text should be spaced appropriately for ease of reading. [Chandra, et al. (2003), 2.4.12]
- A highly legible typeface enables the user to quickly and accurately identify each character. The FAA Human Factors Design Standard for Acquisition of Commercial-off-the-shelf Subsystems, Non-Developmental Items, and Developmental Systems (DOT/FAA/CT-03/05) recommends the following: [Chandra, et al. (2003), 2.4.10]

    Upper case text should be used sparingly. Upper case text is appropriate for single words, but should be avoided for continuous text. (HFDS 8.2.5.8.2)

(a) For continuous text (e.g., sentences and paragraphs), use mixed upper and lower case characters. (HFDS 8.2.5.8.4)

(b) Use serif fonts for continuous text if the resolution is high enough not to distort the serifs (small cross strokes at the end of the main stroke of the letter). (HFDS 8.2.5.7.5)

Sans serif fonts should be used for small text and low resolution displays. (HFDS 8.2.5.7.6)

For optimum legibility, character contrast should be between 6:1 and 10:1. Lower contrasts may diminish legibility, and higher contrasts may case visual discomfort (HFDS 8.2.5.6.12)

Characters stroke width should be 10 to 12% of character height. (HFDS 8.2.5.6.14)

- The FAA *Human Factors Design Standard for Acquisition of Commercial-off-the-shelf Subsystems, Non-Developmental Items, and Developmental Systems* (DOT/FAA/CT-03/05) provides the following recommendations regarding the typeface size and width: [Chandra, et al. (2003), 2.4.11]
    - The minimum character height should be 16 minutes of arc (5 millirad). For practical purposes, this requires a minimum typeface height of 1/200 of the viewing distance. (DOT/FAA/CT-03/05, 8.2.5.6.6)
    - The preferred character height is 20 to 22 minutes of arc (approximately 6 millirad). For practical purposes, this translates into a typeface height of 1/167 of the viewing distance. (DOT/FAA/CT-03/05, 8.2.5.6.5)
    - The ratio of character height to width should be:
        At least 1:0.7 to 1:0.9 for equally spaced characters and when lines of 80 or fewer characters are used.
        At least 1:0.5 if more than 80 characters per line are used.
        As much as 1:1 for inherently wide characters such as "M" and "W" when proportionally spaced characters are used.

    If these guidelines are not met, there should be a sound basis for deviation.

- In order to make text easily readable, the FAA *Human Factors Design Standard for Acquisition of Commercial-off-the-shelf Subsystems, Non-Developmental Items, and Developmental Systems* recommends the following:
    - Use a horizontal spacing between characters of at least 10 percent of character height. (DOT/FAA/CT-03/05, 8.2.5.6.1)
    - Use spacing between words of at least one character when using equally spaced characters or the width of the capital letter "N" for proportionally spaced characters. (DOT/FAA/CT-03/05 8.2.5.6.2)
    - Use a vertical spacing between lines of at least two stroke widths or 15 percent of character height, whichever is larger. Vertical spacing begins at the bottom of character descenders (that part which descends below the text line as seen in the lower-case letter "y") and ends at the top of accent marks on upper case characters. (DOT/FAA/CT-03/05 8.2.5.6.3)

2.8 Symbols

- ❖ All symbols shall be depicted in an upright orientation except for those designed to reflect a particular compass orientation. [TSO C-165/RTCA DO-257A, 2.2.1.1]
- Symbols should be distinguishable based on their shape alone, without relying upon secondary cues such as color and text labels. [Chandra, et al. (2003), 2.4.13]
- Symbols should be designed so that they are discriminable when presented on the minimum expected display resolution when viewed from the maximal intended viewing distance. [Chandra, et al. (2003), 2.4.13]

2.9 Graphical Icons

- Graphical icons should be accompanied by brief text labels if their meaning is not obvious. (See also 2.5 Design of Labels) [Chandra, et al. (2003), 2.4.4]
- If graphical icons are used as labels, the meaning of the icon should be obvious.

- Graphical icons should be designed carefully to minimize any necessary training, and to maximize the intuitiveness of the icon for cross-cultural use. [Chandra, et al. (2003), 2.4.4]

2.10 Configuring Display Properties

- If user-interface customization by the end user is supported, the end user should be provided with an easy means by which to reset all customized parameters back to their default values. [Chandra, et al. (2003), 2.4.19]
- The current operating mode and the functionality being configured should be indicated clearly.

2.11 Failure Conditions

- ❖ The equipment, systems, and installations whose functioning is required by this subchapter, must be designed to ensure that they perform their intended functions under any foreseeable operating condition. [14 CFR §§ 23.1309(b)(1), 25.1309(a), 27.1309(a), 29.1309(a)]
  Related Policy: 14 CFR § 23.1309(b)(1) is worded slightly differently.
- ❖ The airplane systems and associated components, considered separately and in relation to other systems, must be designed so that: [14 CFR §§ 23.1309(b)(2), 25.1309(b)]
  1) The occurrence of any failure condition which would prevent the continued safe flight and landing of the airplane is extremely improbable, and
  2) The occurrence of any other failure conditions which would reduce the capability of the airplane or the ability of the crew to cope with adverse operating conditions is improbable.
  Related Policy: 14 CFR §§ 27.1309(b) and 29.1309(b) are worded slightly differently.
- ❖ Compliance with the requirements of paragraph (b) of this section must be shown by analysis, and where necessary, by appropriate ground, flight, or simulator tests. The analysis must consider -- [14 CFR §§ 23.1309(b)(4), 25.1309(d)]
  1) Possible modes of failure, including malfunctions and damage from external sources.
  2) The probability of multiple failures and undetected failures.
  3) The resulting effects on the airplane and occupants, considering the stage of flight and operating conditions, and
  4) The crew warning cues, corrective action required, and the capability of detecting faults.
- ❖ Any probable failure of the surface moving map shall not degrade the normal operation of other equipment or systems connected to it beyond degradation due to the loss of the surface moving map itself. [TSO C-165/RTCA DO-257A, 3.1.4]
- ❖ The failure of interfaced equipment or systems shall not degrade normal operation of the surface moving map equipment beyond degradation due to the loss of data from the interfaced equipment. [TSO C-165/RTCA DO-257A, 3.1.4]
- ❖ If an application is fully or partially disabled, or is not visible or accessible to the user due to a failure, this loss of function should be clearly indicated to the user with a positive indicator. That is, lack of an indication is not sufficient to declare a failure condition. [AC 120-76A, Section 10.d (2), Chandra, et al. (2003), 2.4.9]

2.12 Update Rate

- ❖ For those elements of the display that are normally in motion, any jitter, jerkiness, or ratcheting effect should neither be distracting nor objectionable. [AC 25-11, 6.e]
- ❖ Movement of map information should be smooth throughout the range of aircraft maneuvers. [TSO C-165/RTCA DO-257A, 2.2.4]
- ❖ Maximum latency of aircraft position data at the time of display update shall be one second, measured from the time the data is received by the display system. [TSO C-165/RTCA DO-257A, 2.2.4]
- ❖ When the display receives a "data not valid" or "reduced performance" (e.g., dead reckoning mode) indication from the source, this condition shall be indicated on the display within one second. [TSO C-165/RTCA DO-257A, 2.2.4]

- If aircraft positioning data are not received by the display for five seconds (i.e., data timeout), this condition shall be indicated to the flight crew. [TSO C-165/RTCA DO-257A, 2.2.4]
- If there is an active flight plan and the flight plan data are not received by the display for 30 seconds, this condition shall be indicated to the flight crew. [TSO C-165/RTCA DO-257A, 2.2.4]
- The display shall update the displayed minimum required information set at least once per second. The minimum required information set for surface moving map displays consists of ownship and runways. [TSO C-165/RTCA DO-257A, 2.2.4]

2.13 Responsiveness
- The display shall respond to operator control inputs within 500 msec. [TSO C-165/RTCA DO-257A, 2.2.4]
- It is desirable to provide a temporary visual cue to indicate that the control operation has been accepted by the system (e.g., hour glass or message). It is recommended that the system respond within 250 msec. [TSO C-165/RTCA DO-257A, 2.2.4]

2.14 Shared Display Considerations
- The minimum flight crew must be established so that it is sufficient for safe operation, considering the workload on individual crewmembers. [14 CFR §§ 23.1523(a), 25.1523(a), 27.1523(a), 29.1523(a)]
- Where information on the shared display is inconsistent, the inconsistency shall be obvious or annunciated, and should not contribute to errors in information interpretation. [TSO C-165/RTCA DO-257A, 2.1.9]
- If information, such as traffic or weather, is with the navigation information on the electronic map display, the projection, the directional orientation and the map range should be consistent among the different information sets. [TSO C-165/RTCA DO-257A, 2.1.9; Chandra, et al. (2003), 6.2.8; SAE ARP 5898, 8.3.5]
- Symbols and colors used for one purpose in one information set should not be used for another purpose within another information set. [TSO C-165/RTCA DO-257A, 2.1.9]
- Deselection of shared information (e.g., weather, terrain, etc.) should be possible to declutter the display or enhance readability. [TSO C-165/RTCA DO-257A, 2.1.9]

3 Surface-Moving-Map Display Elements

3.1 Databases

- ❖ If the airport map database is separate from the navigation information database, the surface moving map shall provide a means to identify the database version, and/or date, and/or valid operating period. [TSO C-165/RTCA DO-257A, 2.3.5]
- ❖ The display shall indicate if any data is not yet effective or is out of date. [TSO C-165/RTCA DO-257A, 2.2.5]
- ❖ There should be a required pilot action acknowledging an expired database. [TSO C-165/RTCA DO-257A, 2.2.5]
- ❖ WGS-84 position reference system or an equivalent earth reference model shall be used for all displayed data. (Reference RTCA DO-236A and ICAO Annex 15). [TSO C-165/RTCA DO-257A, 2.2.5]
- ❖ The process of updating aerodrome databases shall meet the standards specified in RTCA DO-200A/EUROCAE ED-76. [TSO C-165/RTCA DO-257A, 2.3.5]

3.2 Accuracy

- ❖ All displayed symbols and graphics shall be positioned (i.e., drawn or rendered) accurately relative to one another such that placement errors are less than .013 inches on the map depiction or 1% of the shortest axis (i.e., horizontal and vertical dimension) of the map depiction, and orientation errors are less than 3° with respect to the values provided by the position and database sources. [TSO C-165/RTCA DO-257A, 2.2.1]
- ❖ The display shall provide an indication if the accuracy implied by the display is better than the level supported by the total system accuracy. [TSO C-165/RTCA DO-257A, 2.3.1]
- ❖ The total system accuracy shall be sufficient for the intended operation, and shall not exceed 100 meters (95%). The installed system should be evaluated to confirm compliance with the requirement above. [TSO C-165/RTCA DO-257A, 3.2.3]
- ❖ The inaccuracies in the depiction of ownship position should be indicated by depicting a "circle of uncertainty" around the aircraft symbol. The radius of the circle should consider feature placement standards of the originating charting agency and errors introduced by the processing steps. It is recommended that the radius indicate a 2-sigma (95%) confidence level based on a numerical analysis of the inherent errors. Accuracy is also affected by the position sensor. If a position source other than GNSS is used, the position error inherent in the position sensor system must be taken into account and a corresponding increase of the radius of the circle of uncertainty may be required. [TSO C-165/RTCA DO-257A, F.3]
- ❖ It is recommended that manufacturers include text similar to the following in the user manual and/or on a product identification screen: "Note: Discrepancies [of up to Xnm] in the placement of airport and navigational aid symbols are known to exist in the source material. This product is not intended *for navigation guidance.*" [TSO C-165/RTCA DO-257A, F.3]
- ❖ The aircraft position sensor horizontal positional accuracy for runways shall be less than 36m. [TSO C-165/RTCA DO-257A, 2.3.1.1.1]
- ❖ The aerodrome total database accuracy for runways shall be 43m or less. [TSO C-165/RTCA DO-257A, 2.3.1.1.1]
- ❖ The aircraft position sensor horizontal positional accuracy for taxiways shall be less than 36m. [TSO C-165/RTCA DO-257A, 2.3.1.1.2]
- ❖ The aerodrome total database accuracy for taxiways shall be 65m or less. [TSO C-165/RTCA DO-257A, 2.3.1.1.2]
- ❖ If runway markings (e.g., runway centerline) are provided they should be depicted in their correct relative position. [TSO C-165/RTCA DO-257A, 2.3.2]
- ❑ The ownship symbol should only be displayed on maps or charts that are georeferenced and to scale. [Chandra, et al. (2003), 6.2.10]
- ❑ The range of display zoom levels should be compatible with the position accuracy of the ownship symbol. [Chandra, et al. (2003), 6.2.10]

- Text in the pilot's guide and airplane flight manual should document the inaccuracies in the presentation of ownship position and which part of the ownship symbol corresponds to ownship's actual position.
- Loss of ownship positional information should be indicated clearly and immediately.
- If hold lines are provided they should be depicted in their correct relative position.
- If traffic symbols are displayed, text in the pilot's guide and airplane flight manual should document the inaccuracies in the presentation of traffic position and which part of the traffic symbol corresponds to the aircraft's actual position.
- All traffic symbols should be positioned on the display in their appropriate location representative of their actual range.

3.3 Ownship
- The surface map display shall contain a symbol representing the location of ownship. [TSO C-165/RTCA DO-257A, 2.3.1.2]
- The ownship symbol shall be unobstructed. [TSO C-165/RTCA DO-257A, 2.2.1.1]
- If directional data is available, the ownship symbol should indicate directionality. [TSO C-165/RTCA DO-257A, 2.3.1.2]
- If direction/track is not available, the ownship symbol shall not imply directionality. [TSO C-165/RTCA DO-257A, 2.3.1.2]
- If ownship directionality information becomes unusable then this condition should be indicated on the surface map display. [TSO C-165/RTCA DO-257A, 2.3.1.2]
- If the ownship symbol is directional, the front of the symbol that conveys directionality (e.g., apex of a chevron or nose of the aircraft if using an aircraft icon) should correspond to the aircraft location. [TSO C-165/RTCA DO-257A, 2.3.1.2]
- If the ownship symbol is non-directional, the aircraft location should correspond to the center of the non-directional symbol. [TSO C-165/RTCA DO-257A, 2.3.1.2]
- The ownship symbol should be distinct from all other symbology.

3.4 Runways
- The capability shall exist to depict runways. [TSO C-165/RTCA DO-257A, 2.3.1.1.1]
- The depiction of runways shall be distinctive from all other symbology. [TSO C-165/RTCA DO-257A, 2.3.1.1.1]
- With the exception of instances where two or more runways intersect, each runway should be depicted as a contiguous area (i.e., an unbroken rectangle). [TSO C-165/RTCA DO-257A, 2.3.1.1.1]
- Runways should be depicted as filled areas, rather than outlined areas. [TSO C-165/RTCA DO-257A, 2.3.1.1.1]
- When two or more runways intersect, the edges of the intersecting runways should not be drawn through the intersection.
- Runways should be distinguished from other display elements along a dimension other than color. (See also Section 2.1 Use of Color)
- Runways should be depicted with thick solid lines rather than dashed lines so that they will be more salient.

3.5 Taxiways
- The capability should exist to depict taxiways. [TSO C-165/RTCA DO-257A, 2.3.1.1.2]
- Taxiways should be depicted as filled areas, rather than outlined areas. [TSO C-165/RTCA DO-257A, 2.3.1.1.2]
- When taxiways intersect runways, the depiction of runways should be given precedence at the intersection.

- When two or more taxiways intersect, the edges of the intersecting taxiways should not be drawn through the intersection.
- Taxiways should be clearly depicted through ramp areas.

3.6 Runway/Taxiway Identifiers
- ❖ The runway identifiers shall be available for depiction on the display, if available. [TSO C-165/RTCA DO-257A, 2.3.2]
- ❖ If taxiways are depicted then the taxiway identifiers should be available for depiction on the display, if available. [TSO C-165/RTCA DO-257A, 2.3.2]
- ❖ Runway identifiers should be distinguishable from the depiction of runway markings. [TSO C-165/RTCA DO-257A, 2.3.2]
- ❖ At reduced map ranges, at least one identifier should be displayed for any taxiway or runway depicted within the selected map range. [TSO C-165/RTCA DO-257A, 2.3.2]
- ❖ When surface map features are being depicted, the aerodrome designator (e.g., ICAO identifier) or name for the depicted aerodrome should be indicated on the display. [TSO C-165/RTCA DO-257A, 2.3.2]
- Identifiers should remain upright to facilitate readability.
- Runway identifiers when presented should be legible across all map ranges.

3.8 Non-Movement Areas
- The depiction of non-movement areas should be clearly distinguishable from the depiction of movement areas.

3.9 Taxi Route
- ❖ Taxi route information shall be distinguishable from all other map attributes. [TSO C-165/RTCA DO-257A, 2.3.1.1.3]
- ❖ The way taxi routes are depicted in the preview or edit mode shall be distinctive from the depiction of the active taxi route. [TSO C-165/RTCA DO-257A, 2.3.1.1.3]
- ❖ The depiction of taxi routes should not obscure runway or taxiway identifiers. [TSO C-165/RTCA DO-257A, 2.3.1.1.3]

3.10 Prioritization of Map Features
- ❖ To ensure the availability of appropriate information during surface operations, the order of display layer precedence (in case aerodrome features overlap) should be (higher priority layered on top): [TSO C-165/RTCA DO-257A, 2.3.4.1]
  - (a) Ownship symbol
  - (b) Taxi route
  - (c) Runway identifiers
  - (d) Runways
  - (e) Taxiway identifiers
  - (f) Taxiways

3.11 Indicators (Velocity Vectors, Compass Rose, Lubber Line)
- A means to turn the velocity vectors on and off should be provided.
- The units of the horizontal velocity vector should be displayed continuously.
- The units of measurement for the velocity vector should be the same for all displayed traffic and ownship.
- The time value associated with the length of the velocity vector should remain the same when the user zooms in or out of the display.
- Compass rose headings should be labeled, at the least with reference points for north (0°), east (90°), south (180°), and west (270°).

4 Traffic Display

4.1 Traffic Representations
- Each traffic symbol should be positioned at a location representing its relative range and bearing with respect to ownship.
- The traffic symbol should indicate specific directionality, if that data is available and of sufficient quality.
- Surface traffic should be clearly distinguished from airborne traffic. [SAE ARP 5898, 9.4.2.4]
- The crew should be provided a means to select and deselect surface traffic information when appropriate. [SAE ARP 5898, 9.4.2.4]

4.2 Selected Traffic
- A control for turning the target selection feature on/off should be provided. [SAE ARP 5898, 9.8.3.3]
- There should be some means of distinguishing the selected aircraft from other traffic.
- Selected aircraft should remain selected until deselected by the user.
- Unselected vehicles should not obscure the selected vehicle unless they are related to a caution or warning.
- The flight crew should be able to select aircraft targets within the currently displayed range. [SAE ARP 5898, 9.8.3.3]

4.3 Data Blocks/Data Tags
- A data tag should have a clear association with the traffic symbol it references.
- A means should be provided to associate the data block with the traffic symbol.
- The information presented in data tags should be consistent for all aircraft.
- An indication should be provided if any piece of information presented in the data field is not available.
- Traffic identifiers and tags on a display should not obscure each other. The display of tags should be prioritized according to significance to ownship position and route. [SAE ARP 5898, 9.4.3.12]

4.4 Altitude Representations
- A capability to select an altitude band within which to display traffic should be provided to the flight crew.
- If an aircraft creates an alert situation, then that aircraft should be displayed, regardless of the setting of the altitude filter.
- If the crew can filter the display of traffic based on the altitude of the aircraft, the altitude filter setting should be continuously displayed.
- The altitude filter should have a pre-set minimum to ensure that the pilot does not accidentally filter out all traffic.
- Altitude values should be displayed for airborne traffic. If the traffic altitude is not available, an indication that it is not available should be displayed (e.g., NO ALT in the data tag).
- Altitudes for traffic simultaneously displayed should be consistent, all altitudes being displayed either in absolute or relative terms.
- The display should indicate whether absolute or relative altitude is displayed.
- The display should indicate whether traffic is above or below own-ship.
- Altitude values should be inhibited for vehicles on the ground.

5 Functionality

5.1 Map Range and Panning

- ❖ The display shall have the capability of manually changing the map range. [TSO C-165/RTCA DO-257A, 2.2.4]
- ❖ Current map range shall be indicated continuously. [TSO C-165/RTCA DO-257A, 2.2.4]
- ❖ The electronic map display should provide an indication if the map range is smaller (i.e., "zoomed in" closer) than the level supported by the accuracy and resolution of the data. [TSO C-165/RTCA DO-257A, 2.2.1]
- ❖ When using the panning and/or range selection function, an indicator of ownship current position within the overall displayed image should be provided. [TSO C-165/RTCA DO-257A, 2.2.4]
- ❖ If a panning and/or range selection function is available, the equipment should provide the capability to return to an ownship-oriented display with a maximum of two discrete control actions (e.g., two button pushes). [TSO C-165/RTCA DO-257A, 2.2.4]
- ❑ If map range is changed via discrete controls (e.g., buttons or key presses), then separate controls should be provided for increasing and decreasing map range.
- ❑ Range markings within the range arc should be labeled.

5.2 Autozoom

- ❖ If the display is controlling the map range automatically, the mode (e.g., auto map range) should be indicated. [TSO C-165/RTCA DO-257A, 2.2.4]
- ❖ If the automatic map range function is deactivated, the display should maintain the last range scale prior to deactivation until the flight crew manually selects another map range. [TSO C-165/RTCA DO-257A, 2.2.4]
- ❖ If the display is controlling the map range automatically, then the capability shall exist to activate or deactivate the automatic map range. [TSO C-165/RTCA DO-257A, 2.2.4]
- ❑ Changes in map range should be obvious to the user and should not contribute to mode confusion.

5.3 Decluttering

- ❖ The display shall have the capability for manual de-cluttering during operational use. [TSO C-165/RTCA DO-257A, 2.2.1.3]
- ❖ If additional map information has been selected for display, it should be possible to deselect all displayed additional information as a set. [TSO C-165/RTCA DO-257A, 2.2.1.3]
- ❖ It should be possible for the pilot to accomplish this de-clutter function with a single action. [TSO C-165/RTCA DO-257A, 2.2.1.3]
- ❑ The decluttering scheme should be documented in the pilots' guide and in the certification plan.
- ❑ If there is a de-clutter capability, it should not be possible for the pilot to remove safety critical display elements (e.g., terrain, obstructions, or special use airspace) without knowing that they are suppressed. If such information can be de-cluttered, it should not be possible for the pilot to believe that it is not visible because it is not there. [Chandra, et al. (2003), 6.2.11]
- ❑ Managing the display configuration (e.g., scale, orientation, and other options and settings) should not induce significant levels of workload. That is, routine display configuration changes should be minimized. [Chandra, et al. (2003), 6.2.11]
- ❑ Managing the display configuration should not result in a significant increase in head down time nor take attention away from other tasks for extended periods of time.
- ❑ The implementation of any decluttering scheme must be validated to ensure that the display elements are organized in such a way where the information available is usable to the pilot.

5.4 Map Orientation

- ❖ Current map orientation shall be clearly, continuously, and unambiguously indicated (e.g., track-up vs. North-up). [TSO C-165/RTCA DO-257A, 2.2.4]
- ❖ The display shall have the capability to present map information in at least one of the following orientations: actual track-up or heading-up. [TSO C-165/RTCA DO-257A, 2.2.4]

- ❖ If desired track-up orientation is used, the aircraft symbol shall be oriented to actual track or heading. [TSO C-165/RTCA DO-257A, 2.2.4]
- ❖ If the flight crew has selected a display orientation (e.g., track-up), that display orientation should be maintained until an action that requires an orientation change occurs. [TSO C-165/RTCA DO-257A, 2.2.4]
- ❖ If the system is in North-up, the orientation of the map shall be referenced to true North. [TSO C-165/RTCA DO-257A, 2.2.4]
- ❖ In desired track-up orientation, it is recommended that a track extension line that projects the actual track out from the aircraft be displayed. [TSO C-165/RTCA DO-257A, 2.2.4]
- ❖ Consideration should be given to the potential for confusion that could result from presentation of relative directions (e.g., positions of other aircraft on traffic displays) when the display is positioned in an orientation inconsistent with that information. For example, it may be misleading if own aircraft heading is pointed to the top of the display and the display is not aligned with the aircraft longitudinal axis. [AC 120-76A, 10.b(3)]

B.2 REFERENCED CFRS

The following four tables provide the exact wording for sections of 14 CFR §§ 23, 25, 27, and 29, referenced in this document. Regulations for 14 CFR § 23 can be found in Table B-1, 14 CFR § 25 in Table B-2, 14 CFR § 27 in Table B-3, and 14 CFR § 29 in Table B-4. The regulations in each table are ordered according to the section number.

PART 23	
SECTION	**REGULATION**
14 CFR § 23.771(a)	For each pilot compartment, the compartment and its equipment must allow each pilot to perform his duties without unreasonable concentration or fatigue.
14 CFR § 23.773(a)(1)	Each pilot compartment must be arranged with sufficiently extensive, clear and undistorted view to enable the pilot to safely taxi, takeoff, approach, land, and perform any maneuvers within the operating limitations of the airplane.
14 CFR § 23.773(a)(2)	Each pilot compartment must be free from glare and reflections that could interfere with the pilot's vision.
14 CFR § 23.777(a)	Each cockpit control must be located and (except where its function is obvious) identified to provide convenient operation and to prevent confusion and inadvertent operation.
14 CFR § 23.777(b)	The controls must be located and arranged so that the pilot, when seated, has full and unrestricted movement of each control without interference from either his clothing or the cockpit structure.
14 CFR § 23.1301(b)	Each item of installed equipment must be labeled as to its identification, function, or operating limitations, or any applicable combination of these factors.
14 CFR § 23.1309(b)(1)	The design of each item of equipment, each system, and each installation must be examined separately and in relationship to other airplane systems and installations to determine if the airplane is dependent upon its function for continued safe flight and landing and, for airplanes not limited to VFR conditions, if failure of a system would significantly reduce the capability of the airplane or the ability of the crew to cope with adverse operating conditions. Each item of equipment, each system, and each installation identified by this examination as one upon which the airplane is dependent for proper functioning to ensure continued safe flight and landing, or whose failure would significantly reduce the capability of the airplane or the ability of the crew to cope with adverse operating conditions, must be designed to comply with the following additional requirements: (1) It must perform its intended function under any foreseeable operating condition.
14 CFR § 23.1309(b)(2)	When systems and associated components are considered separately and in relation to other systems— (i) The occurrence of any failure condition that would prevent the continued safe flight and landing of the airplane must be extremely improbable; and (ii) The occurrence of any other failure condition that would significantly reduce the capability of the airplane or the ability of the crew to cope with adverse operating conditions must be improbable.

Table B-1. 14 CFR § 23.

\multicolumn{2}{c}{PART 23}
SECTION	REGULATION
14 CFR § 23.1309(b)(3)	Warning information must be provided to alert the crew to unsafe system operating conditions and to enable them to take appropriate corrective action. Systems, controls, and associated monitoring and warning means must be designed to minimize crew errors that could create additional hazards.
14 CFR § 23.1309(b)(4)	Compliance with the requirements of paragraph (b)(2) of this section may be shown by analysis and, where necessary, by appropriate ground, flight, or simulator tests. The analysis must consider— (i) Possible modes of failure, including malfunctions and damage from external sources; (ii) The probability of multiple failures, and the probability of undetected faults.; (iii) The resulting effects on the airplane and occupants, considering the stage of flight and operating conditions; and (iv) The crew warning cues, corrective action required, and the crew's capability of determining faults.
14 CFR § 23.1321(a)	Each flight, navigation, and powerplant instrument for use by any required pilot during takeoff, initial climb, final approach, and landing must be located so that any pilot seated at the controls can monitor the airplane's flight path and these instruments with minimum head and eye movement. The powerplant instruments for these flight conditions are those needed to set power within powerplant limitations.
14 CFR § 23.1321(e)	If a visual indicator is provided to indicate malfunction of an instrument, it must be effective under all probable cockpit lighting conditions.
14 CFR § 23.1322	If warning, caution, or advisory lights are installed in the cockpit, they must, unless otherwise approved by the Administrator, be— (a) Red, for warning lights (lights indicating a hazard which may require immediate corrective action); (b) Amber, for caution lights (lights indicating the possible need for future corrective action); (c) Green, for safe operation lights; and (d) Any other color, including white, for lights not described in paragraphs (a) through (c) of this section, provided the color differs sufficiently from the colors prescribed in paragraphs (a) through (c) of this section to avoid possible confusion. (e) Effective under all probable cockpit lighting conditions.
14 CFR § 23.1381(a)(1)	The instrument lights must make each instrument and control easily readable and discernible.
14 CFR § 23.1523(a)	The minimum flight crew must be established so that it is sufficient for safe operation, considering the workload on individual crewmembers.
14 CFR § 23.1543(b)	For each instrument, each arc and line must be wide enough and located to be clearly visible to the pilot.
14 CFR § 23.1555(a)	Each cockpit control, other than primary flight controls and simple push button type starter switches, must be plainly marked as to its function and method of operation.

Table B-1. 14 CFR § 23. (continued)

PART 25	
SECTION	REGULATION
14 CFR § 25.771(a)	The equipment must allow each flight crew member to perform their duties without unreasonable concentration or fatigue.
14 CFR § 25.773(a)(1)	Each pilot compartment must be arranged to give the pilots a sufficiently extensive, clear, and undistorted view, to enable them to safely perform any maneuvers within the operating limitations of the airplane, including taxiing takeoff, approach, and landing.
14 CFR § 25.773(a)(2)	Each pilot compartment must be free of glare and reflection that could interfere with the normal duties of the minimum flight crew.
14 CFR § 25.777(a)	Each cockpit control must be located to provide convenient operation and to prevent confusion and inadvertent operation.
14 CFR § 25.777(c)	The controls must be located and arranged, with respect to the pilot's seats, so that there is full and unrestricted movement of each control without interference from the cockpit structure or the clothing of the minimum flight crew when any member of this flight crew, from 5'2" to 6'3" in height, is seated with the seat belt and shoulder harness fastened.
14 CFR § 25.1301(b)	Each item of installed equipment must be labeled as to its identification, function, or operating limitations, or any applicable combination of these factors.
14 CFR § 25.1309(b)	The airplane systems and associated components, considered separately and in relation to other systems, must be designed so that: 1) The occurrence of any failure condition which would prevent the continued safe flight and landing of the airplane is extremely improbable, and 2) The occurrence of any other failure conditions which would reduce the capability of the airplane or the ability of the crew to cope with adverse operating conditions is improbable.
14 CFR § 25.1309(c)	Warning information must be provided to alert the crew to unsafe system operating conditions, and to enable them to take appropriate corrective action. Systems, controls, and associated monitoring and warning means must be designed to minimize crew errors which could create additional hazards.
14 CFR § 25.1309(d)	Compliance with the requirements of paragraph (b) of this section must be shown by analysis, and where necessary, by appropriate ground, flight, or simulator tests. The analysis must consider -- 1) Possible modes of failure, including malfunctions and damage from external sources. 2) The probability of multiple failures and undetected failures. 3) The resulting effects on the airplane and occupants, considering the stage of flight and operating conditions, and 4) The crew warning cues, corrective action required, and the capability of detecting faults.
14 CFR § 25.1321(a)	Each flight, navigation, and powerplant instrument for use by any pilot must be plainly visible to him from his station with the minimum practicable deviation from his normal position and line of vision when he is looking forward along the flight path.

Table B-2. 14 CFR § 25.

PART 25	
SECTION	REGULATION
14 CFR § 25.1321(e)	If a visual indicator is provided to indicate malfunction of an instrument, it must be effective under all probable cockpit lighting conditions.
14 CFR § 25.1322	If warning, caution or advisory lights are installed in the cockpit, they must, unless otherwise approved by the Administrator, be— (a) Red, for warning lights (lights indicating a hazard which may require immediate corrective action); (b) Amber, for caution lights (lights indicating the possible need for future corrective action); (c) Green, for safe operation lights; and (d) Any other color, including white, for lights not described in paragraphs (a) through (c) of this section, provided the color differs sufficiently from the colors prescribed in paragraphs (a) through (c) of this section to avoid possible confusion.
14 CFR § 25.1381(a)(1)	The instrument lights must provide sufficient illumination to make each instrument, switch and other device necessary for safe operation easily readable unless sufficient illumination is available from another source.
14 CFR § 25.1523(a)	The minimum flight crew must be established so that it is sufficient for safe operation, considering the workload on individual crewmembers.
14 CFR § 25.1543(b)	For each instrument, each instrument marking must be clearly visible to the appropriate crewmember.
14 CFR § 25.1555(a)	Each cockpit control, other than primary flight controls and controls whose function is obvious, must be plainly marked as to its function and method of operation.

Table B-2. 14 CFR § 25. (continued)

PART 27	
SECTION	REGULATION
14 CFR § 27.771(a)	For each pilot compartment, the compartment and its equipment must allow each pilot to perform his duties without unreasonable concentration or fatigue.
14 CFR § 27.773(a)(1)	Each pilot compartment must be free from glare and reflections that could interfere with the pilot's view and designed so that each pilot's view is sufficiently extensive, clear, and undistorted for safe operation.
14 CFR § 27.777(a)	Each cockpit control must be located to provide convenient operation and to prevent confusion and inadvertent operation.
14 CFR § 27.777(b)	Cockpit controls must be located and arranged with respect to the pilots' seats so that there is full and unrestricted movement of each control without interference from the cockpit structure or the pilot's clothing when pilots from 5'2"; to 6'0" in height are seated.
14 CFR § 27.1301(b)	Each item of installed equipment must be labeled as to its identification, function, or operating limitations, or any applicable combination of these factors.
14 CFR § 27.1309(b)	The equipment, systems, and installations of a multiengine rotorcraft must be designed to prevent hazards to the rotorcraft in the event of a probable malfunction or failure.
14 CFR § 27.1321(a)	Each flight, navigation, and powerplant instrument for use by any pilot must be easily visible to him.
14 CFR § 27.1321(d)	If a visual indicator is provided to indicate malfunction of an instrument, it must be effective under all probable cockpit lighting conditions.
14 CFR § 27.1322	If warning, caution or advisory lights are installed in the cockpit, they must, unless otherwise approved by the Administrator, be— (a) Red, for warning lights (lights indicating a hazard which may require immediate corrective action); (b) Amber, for caution lights (lights indicating the possible need for future corrective action); (c) Green, for safe operation lights; and (d) Any other color, including white, for lights not described in paragraphs (a) through (c) of this section, provided the color differs sufficiently from the colors prescribed in paragraphs (a) through (c) of this section to avoid possible confusion.
14 CFR § 27.1381(a)(1)	The instrument lights must make each instrument, switch, and other devices for which they are provided easily readable
14 CFR § 27.1523(a)	The minimum flight crew must be established so that it is sufficient for safe operation, considering the workload on individual crewmembers.
14 CFR § 27.1543(b)	For each instrument, each arc and line must be wide enough and located to be clearly visible to the pilot.
14 CFR § 27.1555(a)	Each cockpit control, other than primary flight controls and controls whose function is obvious, must be plainly marked as to its function and method of operation.

Table B-3. 14 CFR § 27.

PART 29	
SECTION	**REGULATION**
14 CFR § 29.771(a)	For each pilot compartment, the compartment and its equipment must allow each pilot to perform his duties without unreasonable concentration or fatigue.
14 CFR § 29.773(a)(1)	Each pilot compartment must be arranged to give the pilots a sufficiently extensive, clear, and undistorted view for safe operation.
14 CFR § 29.773(a)(2)	Each pilot compartment must be free of glare and reflection that could interfere with the pilot's view.
14 CFR § 29.777(a)	Each cockpit control must be located to provide convenient operation and to prevent confusion and inadvertent operation.
14 CFR § 29.777(b)	Cockpit controls must be located and arranged with respect to the pilots' seats so that there is full and unrestricted movement of each control without interference from the cockpit structure or the pilot's clothing when pilots from 5'2"; to 6'0" in height are seated.
14 CFR § 29.1301(b)	Each item of installed equipment must be labeled as to its identification, function, or operating limitations, or any applicable combination of these factors.
14 CFR § 29.1309(b)	The rotorcraft systems and associated components, considered separately and in relation to other systems, must be designed so that (1) For Category B rotorcraft, the equipment, systems, and installations must be designed to prevent hazards to the rotorcraft if they malfunction or fail; or (2) For Category A rotorcraft— (i) The occurrence of any failure condition which would prevent the continued safe flight and landing of the rotorcraft is extremely improbable; and (ii) The occurrence of any other failure conditions which would reduce the capability of the rotorcraft or the ability of the crew to cope with adverse operating conditions is improbable.
14 CFR § 29.1309(c)	Warning information must be provided to alert the crew to unsafe system operating conditions, and to enable them to take appropriate corrective action. Systems, controls, and associated monitoring and warning means must be designed to minimize crew errors which could create additional hazards.
14 CFR § 29.1321(a)	Each flight, navigation, and powerplant instrument for use by any pilot must be plainly visible to him from his station with the minimum practicable deviation from his normal position and line of vision when he is looking forward along the flight path.
14 CFR § 29.1321(g)	If a visual indicator is provided to indicate malfunction of an instrument, it must be effective under all probable cockpit lighting conditions.

Table B-4. 14 CFR § 29.

PART 29	
SECTION	REGULATION
14 CFR § 29.1322	If warning, caution or advisory lights are installed in the cockpit, they must, unless otherwise approved by the Administrator, be— (a) Red, for warning lights (lights indicating a hazard which may require immediate corrective action); (b) Amber, for caution lights (lights indicating the possible need for future corrective action); (c) Green, for safe operation lights; and (d) Any other color, including white, for lights not described in paragraphs (a) through (c) of this section, provided the color differs sufficiently from the colors prescribed in paragraphs (a) through (c) of this section to avoid possible confusion.
14 CFR § 29.1381(a)(1)	The instrument lights must make each instrument, switch, and other devices for which they are provided easily readable
14 CFR § 29.1523(a)	The minimum flight crew must be established so that it is sufficient for safe operation, considering the workload on individual crewmembers.
14 CFR § 29.1543(b)	For each instrument, each arc and line must be wide enough and located to be clearly visible to the pilot.
14 CFR § 29.1555(a)	Each cockpit control, other than primary flight controls and controls whose function is obvious, must be plainly marked as to its function and method of operation.

Table B-4. 14 CFR § 29. (continued)

Printed in Great Britain
by Amazon